计算机专业"十四五"精品教材

C++语言设计教程

主 编 苏 菲 王 芬 朱 腾

北京希望电子出版社
Beijing Hope Electronic Press
www.bhp.com.cn

内 容 简 介

　　本书是 C++面向对象编程方面的入门书籍，从面向对象的基本概念讲起，首先让读者了解"面向对象"的概念，然后开始讲解 C++的基本语法、数据结构和基本程序结构。此外，本书还讲述了 C++中流和基本输入/输出流的知识。学习 C++，最重要的就是理解基本概念，例如面向对象、数据类型、数组、指针、链表、函数等。本书结合大量实例，分别讲述了这些重要的知识点。

　　本书既可作为大中专院校计算机专业的教材，也可以作为 C++开发人员的参考用书。书中的有些实例来自实际项目，读者可以参考使用。

图书在版编目（CIP）数据

C++语言设计教程 / 苏菲，王芬，朱腾主编. --北京：北京希望电子出版社，2022.1
　　ISBN 978-7-83002-715-5

　　Ⅰ. ①C... Ⅱ. ①苏... ②王... ③朱... Ⅲ. ①C++语言－程序设计－教材 Ⅳ. ①TP312.8

中国版本图书馆 CIP 数据核字（2021）第 237135 号

出版：北京希望电子出版社	封面：赵俊红
地址：北京市海淀区中关村大街 22 号	编辑：付寒冰
中科大厦 A 座 10 层	校对：龙景楠
邮编：100190	开本：787mm×1092mm 1/16
网址：www.bhp.com.cn	印张：18.5
电话：010-82626270	字数：473 千字
传真：010-62543892	印刷：唐山唐文印刷有限公司
经销：各地新华书店	版次：2022 年 1 月 1 版 1 次印刷

定价：59.80 元

前　言

C++是目前世界上应用最为广泛的计算机高级编程语言之一。C++是一种高效实用的程序设计语言，使用它既可以设计面向过程的程序，也可以设计面向对象的程序。面向对象是 C++的一大特色，它实现了类的封装、继承和多态，这些特性使得 C++的代码具有高度的可维护性和重用性。

由于 C++与 C 语言的特殊关系，使得学过 C 语言的读者看到 C++的代码非常亲切，因为 C++的程序结构和 C 语言的程序结构是一致的。本书是基于读者没有学习过 C 语言的假设条件下编写的，但是对于学习过 C 语言的读者同样适用。

学习程序设计最重要的一点就是要学习编程的思想，本书的第 3 章简单介绍了面向对象的思想，这是所有面向对象程序设计语言的基础。从第 4 章起，分别讲述了 C++面向对象的基础知识，并结合具体实例讲述了 C++中的对象、类、构造函数、析构函数、继承等重要概念，希望读者多上机实践，真正理解面向对象的思想。

本书是 C++程序设计的入门书籍，不会着力于深度分析 C++语言的特性，但会结合实例讲述面向对象的基本概念、C++的基本概念，力求使读者能够通过阅读本书初步了解C++程序设计语言，并能利用 C++语言解决简单的问题。

本书的每一章开始都列出了本章的学习的目标，并且每章后面都有针对该章内容的练习题，特别是程序题，读者一定要上机亲自实践。学习程序设计要多读、多练。多读别人的程序，多上机练习。只有学好 C++的基础知识，才能为以后深入学习和运用 C++打下良好的基础。

本书由苏菲（湖北商贸学院人工智能学院）、王芬（山东劳动职业技术学院）和朱腾（广东工贸职业技术学院）担任主编。本书的相关资料和售后服务可扫描本书封底的微信二维码或登录 www.bjzzwh.com 网站下载获得。

本书难免有疏漏和不当之处，敬请各位专家及读者批评指正。

编　者

目　　录

第1章 C++语言基础

本章学习目标

- C++程序的基本结构
- C++的标识符
- C++的基本数据结构
- C++的关键字

任何一种编程语言都有其语言基础知识，包括程序的基本结构、数据类型、关键字、标识符等。任何程序都是按照其语言基础组织起来的代码序列。本章将讲述C++程序的基本结构，以及重要的C++基本数据结构、标识符和关键字等知识。

1.1 C/C++概述

本节首先讲述C/C++的异同。C++是C的一个超集，20世纪80年代，贝尔实验室的本贾尼·斯特劳斯特鲁普（Bjarme Stroustrup）博士及其同事为了克服C语言的一些固有缺点（比如类型检查机制较弱），在C语言的基础上对其进行了改进和扩充，引入了类的概念，那时的C++被称为带类的C（C With Classes）。其后，本贾尼·斯特劳斯特鲁普博士及其同事不断把新的特性增加进来，例如，虚函数(virtual function)、运算符重载（operator overloading）、多重继承（multiple inheritance）等。1998年国际标准化组织ISO（International Standards Organization）正式发布了C++语言的国际标准ISO/IEC:98-14882，随后各软件厂商推出的C++编译器都支持该标准，并进行了不同的拓展。

C++支持面向对象的程序设计方法，特别是在大型的软件开发项目中，C++在软件的可重用性、可扩展性、可维护性、可靠性等方面都具有巨大的优势。然而，C++与C语言的设计原则是不一样的，C是C++的子集，是面向过程的程序设计语言，而C++则是C的超集，是面向对象的程序设计语言。在Effective C++第三版中，C++被描述成由4个子语言组成：

- C子语言。
- 面向对象的C++。
- 泛型编程语言。C++拥有强大的模板功能，使其能在编译期完成许多工作，从而大大提高了运行期的效率。
- STL。C++标准模板库，其安全性和规范性非常强大。

可以认为，C++是一门独立的程序设计语言，学习C++时可以完全不用学习C语言，并且在大多数场合，C++完全可以代替C语言。然而，在需要直接对硬件进行操作的场合（比如编写硬件驱动程序），仍然需要使用C语言。

1.2 / C++程序的基本结构

下面通过一个例子分析C++程序的基本结构。

实例1-1的程序代码如下，为方便表达，可在每一个代码行的开头都加上行号，这些行号在实际编程中是没有的。

❀ 实例1-1 在屏幕上输出字符。

```
1: #include <iostream.h>
2: void myFunction();

3: int a=1,b=2;
4: main()
5: {
6:   cout<<" Function main:a="<<a<<"\n";
7:   cout<<" Function main:b="<<b<<"\n";
8:   cout<<" Function main:a+b="<<a+b<<"\n";
9:   myFunction();
10: }


11: void myFunction()
12: {
13:   int c=3;
14:   cout<<"Function myFunction:a="<<a<<"\n";
15:   cout<<"Function myFunction:b="<<b<<"\n";
16: cout<<"Function myFunction:c="<<c<<"\n";
17: cout<<"Function myFunction:a+b-c="<<a+b-c<<"\n";
18: }
```

下面来分析一下这个程序的结构。

- **处理命令** 实例1-1的第1行是一条预处理命令，其作用是将头文件iostream.h包含进程序中，以实现程序的基本输入/输出操作。

- **函数** C++程序是由若干个文件组成的，每个文件又由若干个函数组成，函数与函数间是平行的，函数间可以调用。在一个程序中，必须有一个主函数main()。实例1-1的第2行声明了一个自定义函数myFunction()，从程序的第11行到第18行是这个自定义函数的函数体，第9行调用这个函数。第4行到第10行是主函数体，这是每个程序所必需的。

- **输入/输出** 实现与程序内部进行信息交流，几乎每个C++程序都会用到输入/输出。实例1-1的第6~8行以及第14~17行的作用就是在屏幕上输出相关的字符。

- **变量** 几乎所有C++程序都会用到变量，变量在使用前必须先声明。实例1-1的第3行和第13行就分别声明了3个变量a、b和c，并为它们赋予初始值a=1、b=2、c=3。第6、第7、第14和第15行直接将这些变量的值输出到屏幕上。第8和第17行还对这些变量进行了简单运算，并把运算结果输出到屏幕上。

- 语句　语句是组成程序的基本单元，上面提到的函数就是由若干条语句组成的，C++程序中的语句以分号作为结束符。语句有表达式语句、空语句、复合语句、分支语句、循环语句等。

- 表达式　表达式是由常量、变量、函数和运算符组合起来的式子。表达式具有一个值及其类型，它们等于计算表达式所得结果的值和类型。单个常量、变量、函数可以看作表达式的特例。实例1-1的第8和第17行的a+b和a+b-c就是两个表达式。

程序的运行结果如下：

```
Function main:a=1
Function main:b=2
Function main:a+b=3
Function myFunction:a=1
Function myFunction:b=2
Function myFunction:c=3
Function myFunction:a+b-c=0
```

上面分析了一个典型的 C++ 程序的基本结构。下面再给出一个例子，请读者自行分析这个 C++ 程序的基本结构。

实例1-2　通过函数调用，求两个数值加、减、乘、除的值。

```cpp
#include <iostream.h>

void funAdd();
void funSub();
void funMul();
void funDiv();

int a=2,b=4;
main(){
  cout<<"=========Start========="<<"\n";
  cout<<"a ="<<a<<",b ="<<b<<"\n";
  funAdd();
  funSub();
  funMul();
  funDiv();
  cout<<"==========End=========="<<"\n";
```

```
    return 0;
}

void funAdd(){
  cout<<"a + b ="<<a+b<<"\n";
}

void funSub(){
  cout<<"b - a ="<<b-a<<"\n";
}

void funMul(){
  cout<<"a * b ="<<a*b<<"\n";
}

void funDiv(){
  cout<<"b / a ="<<b/a<<"\n";
}
```

程序的运行结果如下：

```
=========Start==========
a = 2,b = 4
a + b = 6
b - a = 2
a * b = 8
b / a = 2
==========End=========
```

实例1-2的基本语法和程序结构与实例1-1类似，有兴趣的读者可试着修改这个例程中的自定义函数funDiv()，例如将b/a变成a/b，结果是什么？

提示

为了使读者能够在运行本书的例子时清晰了解程序的内容，本书的每一个例子都会在输出部分详细描述本例的功能。

1.3 C++的基本数据类型

数据类型是按照被定义变量的性质、表现形式以及占存储空间的多少来划分的。在C++

中，每个变量在使用之前都必须定义其数据类型。

基本数据类型最主要的特点是：其值不可以再分解为其他类型。也就是说，基本数据类型是自我说明的。

C++的基本数据类型包括：

- 整型（int）　用于定义整数对象。
- 字符型（char）　用于定义字符数据。
- 浮点型（float，double）　用于定义实数，包括单精度和双精度类型。
- 空类型（void）　空类型是一个比较特殊的类型，void描述了一个空集，变量不能被定义成void类型，它主要用来声明没有返回值的函数。另外，void也用来声明未确定类型或任意数据类型的指针。

在C++中，可以用一些修饰符来对数据类型做进一步的修饰，包括：

- short　短类型，例如short int表示短整型。
- long　长类型，例如long int表示长整型。
- signed　有符号型，例如signed short int表示有符号短整型。
- unsigned　无符号型，例如unsigned short int表示无符号短整型。

表1-1为C++部分基本数据类型的内存空间占用情况和取值范围。

表1-1　C++基本数据类型

基本数据类型分类	标识方法	内存占用（字节）	取值范围
整型	int	4	-2147483648~2147483647
短整型	short int	2	-32768~32767
有符号短整型	signed short int	2	-32768~32767
无符号短整型	unsigned short int	2	0~65535
长整型	long int	4	-2147483648~2147483647
有符号长整型	signed long int	4	-2147483648~2147483647
无符号长整型	unsigned long int	4	0~4294967295
单精度浮点型	float	4	-3.4×10^{38}~3.4×10^{38}
双精度浮点型	double	8	-1.7×10^{308}~1.7×10^{308}
字符型	char	1	-128~127
有符号字符型	signed char	1	-128~127
无符号字符型	unsigned char	1	0~255

下面举例说明不同数据类型的不同取值范围。

实例1-3　不同数据类型的不同取值范围。

```
#include <iostream.h>


void myFunShort();

void myFunInt();

void myFunStart();


void myFunShortInt(){
```

```
        short int a,b,c;
        a = 32766;
        b = a + 1;
        c = a + 2;
        cout<<"b ="<<b<<"\n";
        cout<<"c ="<<c<<"\n";
}

void myFunInt(){
        int a,b,c;
        a = 32767;
        b = a + 1;
        c = a + 2;
        cout<<"b ="<<b<<"\n";
        cout<<"c ="<<c<<"\n";
}

void myFunStart(){
        cout<<"*****************"<<"\n";
}

main(){
        myFunStart();
        myFunShortInt();
        cout<<"\n";
        myFunInt();
        myFunStart();
}
```

程序运行结果如下：

```
*****************
 b = 32767
 c = -32768


 b = 32768
 c = 32769
*****************
```

上面的例子可以说明以下几个问题：

（1）不同的数据类型具有不同的取值范围。在自定义函数myFunShortInt中定义了short int类型的变量a、b和c，由于short int类型的取值范围是-32 768~32 767，因此给变量a赋值为32 766，然后把变量b的值设为a+1，变量c的值设为a+2。通过程序运行的结果可以看到，变量b的值是32 767，并没有超过short int的取值范围，而变量c的值应该是32 768，但这个值已经超出了short int的取值范围，所以实际得到变量c的值并不是32 768，而是-32 768。

在自定义函数myFunInt中定义了int型的变量a、b和c，由于int的取值范围是-2 147 483 648~2 147 483 647，给变量a、b、c赋予与myFunShortInt相同的值，所以这个函数计算变量c的值就是32 769，证明在这个函数中，变量c并没有超出int的取值范围。

（2）主函数并没有做具体的计算工作，而是调用不同功能的函数。在一个程序中，绝大部分具体目标的实现都是通过不同的功能函数来实现的，主函数一般不做这类工作。

（3）变量具有作用域，正如在函数myFunShortInt和myFunInt中都定义了变量a、b和c，并且在这两个函数中这些变量的类型不一致，通过程序运行的结果可以看到，它们之间互不影响，也就是说3个变量都只在自己所在的函数中有效，离开所在的函数就不存在了，这种变量称为局部变量。在实例1-1和实例1-2中，也定义了变量a和b，但这两个变量在程序文件的主函数和所有自定义函数中均有效，这种变量称为全局变量。

下面再来看一个较为复杂的例子。

实例1-4　根据输入数值判断要调用的函数。

```cpp
#include <iostream.h>

void myFun1();

void myFun2();

void myFunStar();


short int a,b;
void myFun1(){
    cout<<"a - b ="<<a-b<<"\n";
 }


void myFun2(){
    cout<<"b - a ="<<b-a<<"\n";
 }


void myFunStar(){
    cout<<"***************"<<"\n";
 }
```

```
main(){
    cout<<"a =";
    cin>>a;
    cout<<"b =";
    cin>>b;
    myFunStar();
    if (a > b)
        myFun1();
    else
        myFun2();
    myFunStar();
}
```

程序运行结果如下：

```
a = 1
b = 2
***************
  b - a = 1
***************
```

这个例子相对于之前的程序要复杂一些，因为它增加了一个条件判断语句if…else，这个语句的作用是判断表达式的条件是否成立，如果条件成立则执行if后面的语句，否则执行else后面的语句。

关于if…else结构，会在后面的章节中详细讲解，现在只须知道它的意思即可。另外，这个例子还使用了cin来读取用户从键盘输入的数值。读者请自行运行这个程序，并分析程序运行的结果。

1.4 标识符

所谓标识符，就是指程序员所起的名字，用来标识变量、常量、函数、数组、类、对象等。C++中有效的标识符构成规则如下：

（1）第一个字符必须是字母或下划线。

（2）从第二个字符开始，可以是字母、数字或下划线。

（3）标识符区分大小写，例如myVar、myvar、MyVar、myvAr均代表不同的变量。

（4）标识符不能与C++的关键字重名。例如，不能将变量命名为float、int、for、char、short、while、signed等。

1.5　关键字

关键字是C++保留的标识符，在程序中自定义的标识符不能与关键字重名。每个关键字都有其特殊的含义，表1-2列出了C++的关键字。

表1-2　C++的关键字

asm	auto	bad_cast	bad_typeid
bool	break	case	catch
char	class	const	const_cast
continue	default	delete	do
double	dynamic_cast	else	enum
except	explicit	extern	FALSE
finally	float	for	friend
goto	if	inline	int
long	mutable	namespace	new
operator	private	protected	public
register	reinterpret_cast	return	short
signed	sizeof	static	static_cast
struct	switch	template	this
throw	TRUE	try	type_info
typedef	typeid	typename	union
unsigned	using	virtual	void
volatile	while		

1.6　运算符与表达式

运算符是指对常量或变量进行运算及处理的符号，也是构成表达式的重要元素。C++中具有极其丰富的运算符，这些运算符又具有不同的优先级和结合性，使得C++能够进行各种复杂的数学、工程等运算，同时也使C++变得更加复杂。

下面来学习C++的主要运算符。

1．算术运算符

算术运算符用于各类数值运算，包括加（+）、减（-）、乘（*）、除（/）、求余（或称模运算%）、自增（++）、自减（--）。由运算符连接的表达式称为算术表达式，例如a*2-1、b+c*3等。

2．关系运算符

关系运算符用做比较运算，包括大于（>）、小于（<）、等于（==）、大于或等于（>=）、小于或等于（<=）和不等于（!=）。

3．逻辑运算符

逻辑运算符用于逻辑运算，包括与（&&）、或（||）、非（!）。

4．位运算符

参与位运算的量，按二进制位进行运算，包括位与（&）、位或（|）、位非（~）、位异或（^）、左移（<<）、右移（>>）。

5．赋值运算符

用于对变量或表达式赋值，分为简单赋值（=）、复合算术赋值（+=，-=，*=，/=，%=）和复合位运算赋值（&=，|=，^=，>>=，<<=）等。

6．sizeof()运算符

用于计算存储某种数据类型或变量所需的字节数，其格式为：sizeof(<数据类型>) 或 sizeof(<变量名>)。

7．表达式

表达式是由常量、变量、函数和运算符组合起来的式子。表达式有值及其类型，它们等于计算表达式所得结果的值和类型。表达式求值规则按运算符的优先级和结合性规定的顺序进行。单个常量、变量、函数，可以看作表达式的特例。

下面来看几个C++中经常遇到的也是很重要的问题。

（1）运算符的优先级。在表达式中，优先级较高的先于优先级较低的进行运算，而在一个运算量两侧的运算符优先级相同时，则按运算符的结合性所规定的结合方向处理。

（2）运算符的结合性。C++中各运算符的结合性分为两种，即左结合性（自左至右）和右结合性（自右至左）。例如，算术运算符的结合性是自左至右，即先左后右。比如有表达式a-b+c，则b应先与－号结合，执行a-b运算，然后再执行+c运算，这种自左至右的结合方向就称为左结合性，而自右至左的结合方向称为右结合性。

最典型的右结合性运算就是赋值运算。例如a=b=c，由于赋值运算符（=）的右结合性，应先执行b=c，再执行a=(b=c)运算。

（3）自增、自减运算符。自增1运算符记为++，其功能是使变量的值自增1。自减1运算符记为--，其功能是使变量值自减1。自增1和自减1运算符均为单目运算，它们都具有右结合性。可以有以下几种形式：

- ++i i自增1后再参与其他运算。
- --i i自减1后再参与其他运算。
- i++ i参与运算后，i的值再自增1。
- i-- i参与运算后，i的值再自减1。

在理解和使用上容易出错的就是i++和i--。特别是当它们出现在较复杂的表达式或语句中时，常常难于弄清，因此应仔细分析。

实例1-5是一个自增1和自减1的例子。

实例1-5 自增1和自减1运算。

```
1:#include <iostream.h>
2:
3:main(){
4:   short int a = 8;
```

```
 5:    cout<<"***********************"<<"\n";
 6:    cout<<"++a   ="<<++a<<"\n";
 7:    cout<<"--a   ="<<--a<<"\n";
 8:    cout<<"a++   ="<<a++<<"\n";
 9:    cout<<"a--   ="<<a--<<"\n";
10:    cout<<"-a++ ="<<-a++<<"\n";
11:    cout<<"-a-- ="<<-a--<<"\n";
12:    cout<<"***********************"<<"\n";
13: }
```

程序运行结果如下：

```
***********************
++a  = 9
--a  = 8
a++  = 8
a--  = 9
-a++ = -8
-a-- = -9
***********************
```

在实例1-5中，a的初值为8，第6行a加1后输出为9。第7行减1后输出，故为8；第8行输出a为8之后，再加1故为9；第9行输出a为9之后，再减1故为8；第10行输出-8之后，再加1故为9；第11行输出-9之后，再减1故为8。

实例1-6是一个较为复杂的自增和自减1的例子。

实例1-6 较复杂的自增和自减1的例子。

```
#include <iostream.h>

main(){
  int i=5,j=5,p,q;
  cout<<"***********************"<<"\n";
  p=(i++)+(i++)+(i++);
  q=(++j)+(++j)+(++j);
  cout<<"p ="<<p<<"\n";
  cout<<"q ="<<q<<"\n";
  cout<<"i ="<<i<<"\n";
  cout<<"j ="<<j<<"\n";
  cout<<"***********************"<<"\n";
}
```

请读者自行分析这个例子，关键是要牢记自增和自减的先后顺序。

1.7 常量

对于基本数据类型量，按其取值是否可改变又分为常量和变量两种。其值不可以改变的称为常量，可以改变的称为变量。

在C语言中，可以用#define来定义常量，这种常量称为宏常量。C++除了#define外，还可以用const来定义常量，这种常量称为const常量。

提示

常量其实是一种特殊的变量，特殊在它是值不变的变量。

下面是一个使用字符常量的例子。

实例1-7   使用字符常量。

```cpp
#include <iostream.h>
#include <string.h>

void myFun();
void myFunStart();

const char C1='a';
const char C2='b';

short int a;

void myFun(){
    short int b;
    cout<<"a=";
    cin>>a;
    cout<<"b=";
    cin>>b;
    if (a>b)
      cout<<C1;
    else if (a<b)
      cout<<C2;
    cout<<"\n";
}
```

```
void myFunStart(){
    cout<<"*************"<<"\n";
}


main(){
    myFunStart();
    myFun();
    myFunStart();
}
```

程序的运行结果如下：

```
*************
   a= 1
   b= 2
   b
*************
```

这个例子比较简单，自定义函数myFun用来判断用户输入的全局变量a和局部变量b的大小。如果a比b大，则输出字符串常量C1代表的字符a；如果b比a大，则输出字符串常量C2代表的字符b。

1.8 变量

变量，顾名思义就是其值可变的量。使用之前，变量必须先声明，否则程序在编译时编译器会报错而使程序无法通过编译。在前面的几个例子中，我们已经看到了变量的定义和使用方法，下面再通过一个例子来进一步了解变量的声明和使用方法。

需要指出的是，现在在例子中使用的变量类型都是基本类型，一些复杂的变量将在后续的章节中陆续介绍。

实例1-8　求100以内的偶数和。

```
#include <iostream.h>
#include <math.h>


void funSumEven();
void funStar();
#define NUM 100
long evenSum=0;
```

```
void funSumEven(){
    short int a;
    for (a=0;a<NUM+1;a++)  {
        if (a%2==0)
        evenSum += a;
    }
}


void funStar(){
    cout<<"*********************"<<"\n";
}
main(){
    funStar();
    funSumEven();
    cout<<"2+4+6+…+100 ="<<evenSum;
    cout<<"\n";
    funStar();
    return 0;
}
```

求100以内的偶数之和，从数学公式上讲很简单，即2+4+6+…+100，这个例子也比较简单。首先定义一个常量NUM，用来限定取值的范围，然后声明一个全局变量evenSum，用来把偶数做累加。累加过程在自定义函数funSumEven中通过一个for循环来完成。关于for循环的知识将在以后的章节中学习。

看过上面的例子，读者不妨模仿它自己实现一个程序求100以内所有奇数之和。

1.9 枚举类型

枚举类型是将变量的取值范围规定在一个范围之内，不允许超出这一范围。枚举类型在解决实际问题时有着广泛的应用。例如，一星期内只有7天，一年只有12个月等。只能把这些变量声明为整型，而字符型或其他类型显然是不妥当的。为此，C++语言提供了一种称为枚举的类型。在枚举类型的定义中列举出所有可能的取值，被声明为该枚举类型的变量取值，不能超过其定义的范围。

枚举类型定义的一般形式是：

```
enum 枚举名 {
   枚举值表
   };
```

其中，枚举值表里应罗列所有可用的值，这些值也称为枚举元素。

例如：

```
enum weekday {sun, mon, tue, wed, thu, fri, sat};
```

该枚举名为weekday，枚举值共有7个，即一周中的7天。凡被声明为weekday类型变量的取值，都只能是7天中的某一天。

枚举变量可以采用不同的方式进行声明，即先定义后声明，定义的同时加以声明或直接声明。

（1）先定义后声明。

```
enum weekday{sun, mon, tue, wed, thu, fri, sat};
enum weekday a, b, c;
```

（2）定义的同时加以声明。

```
enum weekday{sun, mon, tue, wed, thu, fri, sat}a,b,c;
```

（3）直接声明。

```
enum{sun, mon, tue, wed, thu, fri, sat}a,b,c;
```

枚举类型在使用时有如下规定：

（1）枚举值是常量，而不是变量。不能在程序中用赋值语句再对它赋值。例如对上面定义的枚举weekday中的元素作以下赋值是错误的。

```
sun = 5;
mon = 1;
```

（2）枚举元素本身由系统定义一个表示序号的数值，从0开始顺序定义为0，1，2…比如在weekday中，sun值为0，mon值为1，…sat值为6。

（3）只能把枚举值赋予枚举变量，而不能把元素的数值直接赋予枚举变量。

```
a=sum; b=mon;
```

这是正确的。

```
a=0; b=1;
```

却是错误的。如果一定要把数值赋予枚举变量，则必须用强制类型转换。

```
a = (enum weekday)2;
```

其意义是将序号为2的枚举元素赋予枚举变量a，相当于a=tue。还应说明的是，枚举元素不是字符常量，也不是字符串常量，使用时不要加单引号或者双引号。

实例1-9说明了枚举类型中各枚举元素的序号。

实例1-9　输出枚举元素的序号。

```
#include <iostream.h>

void funDefineEnum();

void funStarStart();

void funStarEnd();
```

```
void funDefineEnum(){
  enum weekday{ sun,mon,tue,wed,thu,fri,sat } a,b,c;

  a=sun;

  b=mon;

  c=tue;
  funStarStart();
  cout<<"a ="<<a<<"\n";
  cout<<"b ="<<b<<"\n";
  cout<<"c ="<<c<<"\n";
  funStarEnd();
}


void funStarStart(){
    cout<<"**********Begin**********"<<"\n";
}


void funStarEnd(){
    cout<<"**********End**********"<<"\n";
}


main(){
  funDefineEnum();
}
```

程序的运行结果如下：

```
**********Begin**********
a = 0
b = 1
c = 2
**********End**********
```

　　这个例子说明了枚举类型中每个枚举元素的序号。另外，程序中的主函数几乎什么都没有做，仅仅调用了自定义函数funDefineEnum()，后者再调用其他函数。事实上，这么写这个例子的目的是让读者更深刻地了解C++中所有具体的事件处理均交由相关的类来完成。关于类的概念是本书的核心，将在后续章节中详细介绍。这些例子并非说明主函数什么都不能做，事实上把具体事件处理过程交给主函数做同样能够达到目的，只是这么做不符合C++的思想。

1.10　上机操作

学习编程最重要的方法就是多动手，亲自试验书中的例子，并根据自己的思考对例子做出修改。在这个过程中，大家会学到很多东西，而且容易对知识深刻地理解和消化。本书中提供了大量的实例，每个实例读者都可以按照自己的思考角度做出调整，所以读者一定要多动脑、多动手。

本书的每一章都会提供一节"上机操作"，目的是为读者提供亲自上机操作的例子。这些例子都比较容易，但是可以通过它们梳理本章学习的知识，使读者能更深入地理解本章的知识点。请读者一定好好思考，好好练习这些例子。

1.10.1　经典的Hello World程序

上面几节已经接触了一些实例程序，有些还相对比较复杂，而下面这个例子看起来就容易多了。

实例1-10　Hello World程序。

```
#include <iostream.h>

main(){
    cout<<"Hello World!"<<"\n";
    return 0;
}
```

这个程序非常简单，只须注意以下几点：

（1）一个程序中只能有一个main()函数，但是可以有多个自定义函数。

（2）\n为换行符。

（3）return 0为程序的返回值。

（4）每条语句必须以分号作为结束符。

读者可以思考：可否将这个简单的程序改为由主函数调用自定义函数来实现目标的相对复杂一点的程序呢？应该怎么修改？其中应注意什么问题？

1.10.2　常量与变量练习

常量与变量的练习在前面学习常量和变量时已经看过几个例子，下面的例子继续帮助读者深化对常量和变量的理解。

实例1-11　求三角形的面积。

```
#include <iostream.h>

void funInput();
void funCalcArea();
```

```
    void funStar();


    const short int CONVALUE=2;


    void funStar(){
        cout<<"*********************************"<<"\n";
    }


    void funInput(){
        float a=0.0; /* 三角形的底 */
        float h=0.0; /* 三角形的高 */

        cout<<" 三角形面积的计算公式为：(底 * 高)/2"<<"\n";
        cout<<" 请输入三角形的底：底 = ";
        cin>>a;
        cout<<" 请输入三角形的高：高 = ";
        cin>>h;
        cout<<" 三角形的面积为：(底 * 高)/2 = "<<(a*h)/CONVALUE<<"\n";
    }


    main(){
        funStar();
        funInput();
        funStar();
    }
```

程序运行结果如下：

```
*********************************
三角形面积的计算公式为：(底 * 高)/2
请输入三角形的底：底 = 5
请输入三角形的高：高 = 6
三角形的面积为：(底 * 高)/2 = 15
*********************************
```

读者可以模仿本例自行编写求平行四边形面积、菱形面积的程序，也可以试着将常量的声明方式由const变为#define。

在实例1-8中，已经学习了如何用C++编程求1～100中所有偶数之和，并且给读者留下一个思考题：如何求1～100中所有奇数之和？相信读者已经写出了用C++编程求奇数和的程序。

　　下面这个例子把求偶数之和与求奇数之和结合起来，通过用户选择来判断程序该做什么计算。

　　实例1-12　求1～100内的奇数和与偶数和。

```
#include <iostream.h>
#include <math.h>

void funSumEven(); /* 偶数和 */
void funSumOdd();  /* 奇数和 */
void funStar();

#define NUM 100
long evenSum=0;
long oddSum=0;

void funSumEven(){
    short int a;
    for (a=0; a<NUM+1; a++)  {
        if (a%2==0)
          evenSum += a;
    }
}

void funSumOdd(){
    short int a;
     for ( a=1;a<=50;a++ )
      oddSum += 2*a-1;
}

void funStar(){
    cout<<"********************"<<"\n";
}
main(){
    short int x;
    cout<<"0: 求 100 以内的偶数和 "<<"\n";
    cout<<"1: 求 100 以内的奇数和 "<<"\n";
    cout<<" 请输入你要做的计算 :"<<"\n";
```

```
    cin>>x;
    funStar();
    if (x==0)  {
      funSumEven();
      cout<<"2+4+6+…+100 ="<<evenSum;
     }
    else if (x==1)  {
      funSumOdd();
      cout<<"1+3+5+…+99 = "<<oddSum;
    }

    cout<<"\n";
    funStar();
    return 0;
  }
```

　　这个例子反映的程序结构与前面看到的例子是一样的，细心的读者会发现，通过自定义函数funSumEven和funSumOdd求偶数之和与求奇数之和时，所采用的算法不一样。这个例子说明：要达到同一个目标，可以有很多种算法，读者在学习中要发散思维，多想几种算法来实现目标，并且亲自上机去验证，算法才是程序的灵魂。

1.10.3　运算符与表达式练习

　　下面的这个例子综合表达了运算符和表达式的应用。

　　实例1-13　运算符和表达式的应用。

```
#include <iostream.h>

void funCalc();
void funStar();

const short int NUMBER=5;
int numSum=0;

void funStar(){
    cout<<"*************************"<<"\n";
}

void funCalc() {
```

```
short int a,x;
cout<<" 请为变量 a 赋值   a = ";
cin>>a;
cout<<"\n";
cout<<"0: 加法运算 +"<<"\n";
cout<<"1: 减法运算 -"<<"\n";
cout<<"2: 乘法运算 *"<<"\n";
cout<<"3: 除法运算 /"<<"\n";
cout<<"4: 求余运算 %"<<"\n";
cout<<" 请选择你要进行的操作: ";
cin>>x;
cout<<"\n";


if (x==0)  {
  numSum = a + NUMBER;
  cout<<a<<" +"<<NUMBER<<" ="<<numSum<<"\n";
 }


if (x==1)  {
  numSum = a - NUMBER;
  cout<<a<<" -"<<NUMBER<<" ="<<numSum<<"\n";
 }


if (x==2)  {
  numSum = a * NUMBER;
  cout<<a<<"*"<<NUMBER<<" ="<<numSum<<"\n";
 }


if (x==3)  {
  numSum = a / NUMBER;
  cout<<a<<"/"<<NUMBER<<" ="<<numSum<<"\n";
 }


if (x==4)  {
  numSum = a % NUMBER;
  cout<<a<<"%"<<NUMBER<<"="<<numSum<<"\n";
```

```
        }
    }


main(){
    funStar();
    funCalc();
    funStar();
}
```

程序运行结果如下：

```
**************************

请为变量 a 赋值  a = 2


0：加法运算 +

1：减法运算 -

2：乘法运算 *

3：除法运算 /

4：求余运算 %

请选择你要进行的操作：4


2 % 5 = 2

**************************
```

1.11 C++开发环境简介

C++有众多的开发环境，例如：Turbo C++、C++ Builder、Visual C++等。其中最为著名的就是微软的Visual C++和Borland的C++ Builder，本书采用的是微软的Visiual C++作为C++实例的开发环境。Visual C++集成开发环境集代码编辑、编译、连接和调试于一体，为编程人员提供了完整而又方便的开发界面。同样，Borland的C++ Builder也是一个集成开发环境，和微软的Visual C++集成开发环境类似。两种C++开发环境都功能强大，读者可根据自己的喜好和要求进行选择。

在众多可视化开发环境中，Visual C++在Windows操作系统下是开发Windows应用程序的最佳选择之一。本书代码是在Visual C++上开发的。

Visual C++ 是Microsoft Visual Studio集成开发环境中一系列开发工具组合之一。Microsoft Visual Studio也随着软硬件的发展，已经有不少版本，典型的有2005、2008、2010、2015、2017、2019等，可以根据需要选择适合自己使用的版本。

选好版本后，安装过程大体都差不多，在此不再详细介绍。

1.12　小结

本章讲述了C++语言的基础知识。学习任何一种编程语言，都要事先学习它的语言规则、数据类型、语句的构成和语法等相关知识。本章目的是让读者初步学习C++的语言基础以及相关的语法知识，为后面章节的学习打下基础。

1.13　习题

一、填空题

1．C语言是面向_____的语言，C++语言是面向_____的语言。

2．在每一个C++程序中，有且只有一个_____函数。

3．变量在使用前必须先_____该变量。

4．_____是组成程序的基本单元，它包括_____、_____、_____等。

5．_____是程序的灵魂。

二、代码阅读题

1．补充完整下面的程序，该程序运行的结果如下：

```
************************
++a  = 4
--a  = 3
a++  = 3
a--  = 4
-a++ = -3
-a-- = -4
************************
```

程序段如下：

```cpp
#include <iostream.h>

main(){
   short int a = 3;
   cout<<"************************"<<"\n";
   cout<<"_____    ="<<_____  <<"\n";
   cout<<"_____    ="<<_____  <<"\n";
   cout<<"_____    ="<<_____  <<"\n";
   cout<<"_____    ="<<_____  <<"\n";
   cout<<"_____    ="<<_____  <<"\n";
```

```
    cout<<"_____    ="<<_____<<"\n";
    cout<<"**************************"<<"\n";
}
```

三、问答题

1．学习C++之前，是不是必须先学习C？

2．怎么才能学好C++？

3．C++可以被描述成4个子语言，请列举这4个子语言。

4．C++是否可以完全替代C？

第2章 C++面向对象基础

本章学习目标

- 类与对象
- 析构函数
- 派生类
- 构造函数
- 类的继承性
- 友元

结构化程序设计的先驱Niklaus Wirth曾经给程序做的定义是：

程序 = 算法 + 数据结构

这种定义针对面向过程的程序，而面向对象程序（OOP，Object Oriented Programming）设计思想是一种基于结构分析，以数据为中心的设计方法，对于OOP，可以做如下的定义：

程序 = 对象 + 消息传递

学习OOP方法，必须学习面向对象中几个重要的概念：对象、类、消息、方法、继承等。下面几节将分别讲述这些内容。理解这些重要概念是学习OOP的基础和先决条件，希望读者细心阅读，深入理解其中的每一个概念，为学习面向对象打好基础。

本章将结合C++具体讲述面向对象的概念，以及这些概念的应用，构造函数、析构函数的作用和使用方法等内容。

2.1 面向对象的思想

传统软件开发方法存在诸如软件重用性差、可维护性差，常常不能真正满足用户需求等问题，而且用结构化方法开发的软件在稳定性、可修改性和代码重用性方面都比较差。

结构化程序设计方法（Structured Programming）围绕过程来展开，它不断地把复杂的问题逐层分解成单独的子问题，直到仅剩下若干个容易实现的子问题过程为止，然后再利用合适的工具来描述各个子问题，从而完成软件开发。这种方法忽视了用户需求，因为用户往往提出功能上的需求变化，按照结构化程序设计方法，可能需要把整个过程重新设计，这对已经开发完成的软件来说可能是灾难性的！

为解决这些问题，C++引入了面向对象的概念。面向对象方法是以认识论为基础，用对象来解释和分析问题，并设计出由对象构成软件系统的方法。面向对象就是面向事情的本身，面向对象方法从对象出发，继而发展出对象、类、消息和集成等概念。

OOP的优点是：符合人们通常的思维方式，软件分析、设计和编码均采用一致的模型，使得软件开发的各个方面均能高度统一。

2.2 / 类和对象

类和对象是面向对象的基础概念，本节将结合实例详细讲述类的概念和对象的概念。

> **提示**
>
> 类和对象是面向对象概念中最基础也是最重要的概念，读者对类和对象的理解一定要深入，要多看实例、多做实验，不能仅仅停留在表面上。对于以前学习过C语言的读者，要更好地理解面向对象的概念。同时本章也是比较理论化的一章，读者要耐心细致地阅读，切忌似懂非懂！

2.2.1 对象

对象就是要研究的事物。在日常生活中，我们每时每刻都在接触不同的对象，例如计算机、桌子、人、汽车、飞机等都是对象。对象都有属性，如果把人看作一个对象，那么人所具有的诸如性别、年龄、身高、体重等都是人这个对象的属性。

把飞机看作是一个对象，那么飞机所具有的制造商、长度、高度、最大飞行速度等都是飞机对象的属性。对于人这个对象，不仅具有属性，还具有行为。例如人吃饭、走路、学习、睡觉等都是人这个对象的行为。飞机起飞、巡航、降落等则都是飞机对象的行为。

属性即为对象的数据，用来描述对象所具有的参数，行为是对象可以进行的活动，即对象可以进行的操作。因此，在面向对象程序设计中，可以概括地把对象用一个等式来描述：

对象 = 数据 + 操作

其中操作被称为"方法"，即加载到对象上的方法，它定义了在对象中如何操作数据。可以说，对象是一个实体，是一个拥有一组数据，并且拥有作用在这组数据上的方法所构成的一个实体。

对象组合了数据和操作，将数据和操作封装到对象中构成一个整体。

2.2.2 类

自然界中具有无数个对象，我们不可能一一描述，但是却可以把性质相同或相似的对象归结为一个类，该类中的每个对象都具有相同的属性和行为。

类即是对象的抽象，而对象则是类的实体化。例如，把人抽象化成一个类，这个类具有相同的属性：姓名、性别、身高、体重等；同时这个类也具有一些行为：吃饭、睡觉、学习、工作、休息等。类与对象是密切相连的，没有脱离对象的类，也没有不依赖于类的对象。

类是一种复杂的数据类型，它是一个将不同类型的数据和与这些数据相关的操作封装在一起的集合。在类中，对类的数据（属性）的操作是通过函数来完成的，这类函数称为成员函数。

在C++中，定义类的基本格式如下：

```
class <类名> {
  public:
    <公有数据成员和成员函数说明部分>
```

```
    private:
        <私有数据成员和成员函数说明部分>
    protected:
        <保护数据成员和成员函数说明部分>
}
<成员函数的实现>
```

定义一个类时以关键字class开头，类名是个字符串，用来命名下面定义的类。花括号内是类的说明，类的成员包括数据成员和成员函数。类中的数据和函数具有不同的访问权限，不同的权限通常使用访问控制修饰符来表示，包括：

- public　公有部分。
- private　私有部分。
- protected　保护性部分。

通常情况下，public部分用来说明公用的成员函数，这类函数可以被后续程序所调用，提供给用户操作该类的接口。private用来说明私有数据，这些数据通常是该类中对象的属性。被private说明的数据是用户无法访问的，通常只有成员函数和友元函数才能访问这部分数据。protected用来说明保护数据，这类数据能被该类中的函数、子类中的函数以及其友元函数访问。

类中函数的具体实现可以在类体中实现，也可以在<成员函数的实现>部分完成。

下面是一个类定义的例子：

```
/*定义类结构*/
class THoman{
    /*公有成员*/
    public:
        void setNO(int NO);
        void setSex(char Sex);
        void outputInfo();
    /*私有成员*/
    private:
        int studentNO,studentSex;
}

/*成员函数的实现*/
/*注意::号是作用域运算符，用来标识成员函数所属的类*/
void THoman::setNO(int NO){
    studentNO=NO;
}
```

```
void Thoman::setSex(char Sex){

    studentSex=Sex;

}


void THoman::outputInfo(){

    cout<<" 学号："<<studentNO<<endl;

    cout<<" 性别："<<studentSex<<endl;

}
```

上面这个类仅简单定义了学生学号和性别，其中公用部分定义了3个函数，前两个函数用来设置在私有部分定义的该类的两个属性，函数outputInfo用来将设置后的信息输出。在这个例子中，把所有成员函数的实现都放在类体的外面进行，实际上成员函数的实现也可以在类体中进行，如下：

```
/* 定义类结构 */
class THoman{

    /* 公有成员，并实现成员函数 */
    public:
        void setNO(int NO)  {

            studentNO=NO;

        }


        void setSex(char Sex)  {

            studentSex=Sex;

        }

        void outputInfo();

        void outputInfo()  {

            cout<<" 学号："<<studentNO<<endl;

            cout<<" 性别："<<studentSex<<endl;

        }


    /* 私有成员 */
    private:
        int studentNO,studentSex;

}
```

这样，便将成员函数的实现放在类体中了。

class为类类型定义的关键字，类体中的public、private和protected关键字的作用已经在前面介绍过，这里不再赘述。需要特别指出的是，当类成员没有显式地规定时，则属于哪一种访问属性呢？默认为private。

下面通过一个简单的例子来说明C++中类的使用。

实例2-1　C++中类的使用。

```cpp
#include <iostream.h>

void myFun();

class TCalc{
    public:
        float add(float x, float y);
        float sub(float x, float y);
        float mul(float x, float y);
        float div(float x, float y);
};

/* 在类体外实现成员函数 */
/* 加法运算 */
float TCalc::add(float x, float y){
    return (x+y);
}


/* 减法运算 */
float TCalc::sub(float x, float y){
    return (x-y);
}


/* 乘法运算 */
float TCalc::mul(float x, float y){
    return (x*y);
}


/* 除法运算 */
float TCalc::div(float x, float y){
    return (x/y);
}
```

```
void myFun(){
    TCalc result;
    float a,b;
    float z;

    cout<<"a=";
    cin>>a;
    cout<<"b=";
    cin>>b;
    cout<<endl;

    cout<<"a+b="<<result.add(a,b)<<endl;
    cout<<"a-b="<<result.sub(a,b)<<endl;
    cout<<"a*b="<<result.mul(a,b)<<endl;

    if (b==0)
        cout<<" 除数不能为 0!"<<endl;
    else
        cout<<"a/b="<<result.div(a,b)<<endl;
}

main(){
    cout<<"*********************************"<<endl;
    cout<<"*        用类实现简单的四则运算        *"<<endl;
    cout<<"*********************************"<<endl;
    myFun();
    cout<<"*********************************"<<endl;
}
```

程序运行结果如下：

```
*********************************
*        用类实现简单的四则运算        *
*********************************
a=109
b=99.9
```

```
a+b=208.9
a-b=9.1
a*b=10889.1
a/b=1.09109
********************************
```

在这个例子中，把简单的数学四则运算写成了一个类TCalc（一般定义类的类名都以大写字母T开头）。这个类只有一个public限制修饰符，表明类中所有成员均能被用户访问。在这个类中，我们定义了4个成员函数：

```
float add(float x, float y);
float sub(float x, float y);
float mul(float x, float y);
float div(float x, float y);
```

这4个函数分别实现了加、减、乘和除法运算。在类体中，只是定义了成员函数，并没有实现。成员函数的实现是放在类体外进行，这时需要使用作用域限制符（::）来表明被实现的函数属于哪个类。实现四则运算的函数都很简单，这里不再赘述。

2.3 构造函数

在C++成员函数中，有两种特殊的函数：构造函数和析构函数。这两种函数在C++类结构中有着特殊和重要的用途，因此将在本节讲述构造函数，下一节讲述析构函数。

类中有类的数据成员，构造函数的作用就是为这些类成员赋初值。构造函数具有以下特点：

（1）构造函数的函数名与类名相同。

（2）构造函数将完成为类的数据成员赋初值和为指针数据成员动态分配存储空间。

（3）构造函数在该类的对象建立时被系统自动调用，不能被用户调用。因此，构造函数必须被说明为该类的公用成员。

（4）构造函数没有返回类型。

C++规定，每个类都必须有构造函数，否则不能使用此类来创建对象。如果没有为该类定义构造函数，那么系统会自动为该类定义一个构造函数，但系统自定义的构造函数仅仅负责创建对象，而不会为成员赋值。如果用户定义了构造函数，那么系统就不会为这个类定义构造函数了。

构造函数可以是无参函数，也可以是有参函数。

实例2-2　定义并使用构造函数。

```
#include <iostream.h>


class TstudentScore{

    public:

        /* 定义构造函数 该函数无形参 */
```

```
        TstudentScore()  {
            /* 完成成员的赋值操作 */
            NO=1000;
            Chinese=100;
            Math=99;
            English=99.5;
        }
        /* 成员函数  功能：输出数据成员值 */
        void coutScore();
    private:
        int NO;
        float Chinese,Math,English;
};


/* 实现成员函数 */
void TstudentScore::coutScore(){
    cout<<"NO="<<NO<<endl;
    cout<<"Chinese="<<Chinese<<endl;
    cout<<"Math="<<Math<<endl;
    cout<<"English="<<English<<endl;
}


main(){
    cout<<"******************************"<<endl;
    cout<<"*         定义构造函数            *"<<endl;
    cout<<"******************************"<<endl;
    TstudentScore Score1;
    Score1.coutScore();
    cout<<"******************************"<<endl;
}
```

程序的运行结果如下：

```
******************************
*         定义构造函数            *
******************************
NO=1000
Chinese=100
```

```
Math=99

English=99.5

****************************
```

在这个例子中，首先定义了一个类TstudentScore，在这个类的public段定义了一个构造函数，以便给TstudentScore类中的私有变量赋值。public段定义的另外一个函数是coutScore，这个函数的目的是将TstudentScore类中定义的私有变量的值输出，该函数在类体外实现。

上面例子中的构造函数是无参函数，构造函数同样可以带有参数。

实例2-3　有参数的构造函数。

```cpp
#include <iostream.h>

class TstudentScore{
    public:
        /* 定义构造函数 该函数有形参 */
        TstudentScore(int result) {
        /* 完成成员的赋值操作 */
            NO=1000;
            Chinese=100;
            Math=99;
            English=99.5;

            if (result==0)  {
              cout<<" 你选择的是求和操作 : "<<endl;
              Result=Chinese+Math+English;
            }
            else if (result==1)  {
              cout<<" 你选择的是求平均值操作 : "<<endl;
              Result=(Chinese+Math+English)/3;
            }
        }
        /* 成员函数 功能 : 输出数据成员值 */
        void coutScore();
    private:
        int NO;
        float Chinese,Math,English;
        float Result;
};
```

```
/* 实现成员函数 */
void TstudentScore::coutScore(){
    cout<<"NO="<<NO<<endl;
    cout<<"Chinese="<<Chinese<<endl;
    cout<<"Math="<<Math<<endl;
    cout<<"English="<<English<<endl;
    cout<<"Result="<<Result<<endl;
}

main(){
    cout<<"*****************************"<<endl;
    cout<<"*          有参构造函数           *"<<endl;
    cout<<"*****************************"<<endl;
    int a;

    cout<<" 请选择计算种类 : "<<endl;
    cout<<"0: 求      和 "<<endl;
    cout<<"1: 求平均值 "<<endl;
    cout<<"a=";
    cin>>a;

    /* 注意：这里这样使用有参构造函数 */
    TstudentScore Score1(a);
    Score1.coutScore();
    cout<<"*****************************"<<endl;
}
```

程序的运行结果如下：

```
*****************************
*          有参构造函数           *
*****************************
请选择计算种类 :
0: 求      和
1: 求平均值
a=0
你选择的是求和操作 :
```

```
NO=1000

Chinese=100

Math=99

English=99.5

Result=298.5

****************************
```

在这个例子中，同样定义了一个名为TstudentScore的类，但是这个类中定义的构造函数与实例2-2不同，前者的构造函数是无参函数，本例的构造函数却是一个有参函数。本实例中的构造函数定义了一个int型变量，该变量的作用是根据用户的选择来执行相应的操作。需要注意的是，在程序中构造这个类的实例时，需要赋予其参数，即格式应为：

```
TstudentScore Score1(a);
```
而下面的格式是错误的：

```
TstudentScore Score1;
```
因为在类TstudentScore中并没有一个无参的构造函数。

一个类中可以有一个或多个构造函数，每个构造函数用来完成不同的任务。

实例2-4　多构造函数的类。

```
#include <iostream.h>


class TstudentScore{
    public:
    /* 定义构造函数1 功能：私有变量赋值 */
    TstudentScore()  {
       /* 完成成员的赋值操作 */
       NO=1000;

       Chinese=100;

       Math=99;

       English=99.5;

       sum=Chinese+Math+English;

    }


    /* 定义构造函数2 功能：求汇总数与平均值 */
    TstudentScore(int a)  {
       /* 完成成员的赋值操作 */
       NO=1000;

       Chinese=100;

       Math=99;

       English=99.5;
```

```
        avg=(Chinese+Math+English)/3;
    }

    /* 成员函数  功能：输出数据成员值 */
    void coutPrivateValue();
    void coutSumResultValue();
    void coutAvgResultValue();

private:
    int NO;
    float Chinese,Math,English;
    float sum,avg;
};

/* 实现成员函数 */
void TstudentScore::coutPrivateValue(){
    cout<<"NO="<<NO<<endl;
    cout<<"Chinese="<<Chinese<<endl;
    cout<<"Math="<<Math<<endl;
    cout<<"English="<<English<<endl<<endl;
}


void TstudentScore::coutSumResultValue(){
    cout<<"Sum="<<sum<<endl;
}


void TstudentScore::coutAvgResultValue(){
    cout<<"Avg="<<avg<<endl;
}


main(){
    cout<<"*****************************"<<endl;
    cout<<"*         定义构造函数           *"<<endl;
    cout<<"*****************************"<<endl;
    TstudentScore Score1;
    TstudentScore Score2(1);
```

```
        Score1.coutPrivateValue();

        Score1.coutSumResultValue();

        Score2.coutAvgResultValue();

        cout<<"******************************"<<endl;

}
```

程序的执行结果如下：

```
******************************
*       定义构造函数          *
******************************
NO=1000
Chinese=100
Math=99
English=99.5


Sum=298.5
Avg=99.5
******************************
```

在这个例子中，类TstudentScore中定义了两个构造函数：TstudentScore()和TstudentScore(int a)。其中，第一个构造函数是无参构造函数，第二个构造函数是有参构造函数，这两个构造函数以此加以区分，称为函数的重载（函数重载的概念将在第6章讲解）。在主程序中，根据构造函数参数的不同而调用不同的构造函数。

2.4 析构函数

析构函数和构造函数一样，也是类中的特殊函数。析构函数的作用是：当对象使用完毕时，系统将自动调用析构函数，以便做对象撤销时的善后工作，例如释放一些动态存储空间等。

析构函数具有如下特点：

（1）析构函数名与类名相同，只是在类名前加一个"~"符号，以区别于构造函数。

（2）析构函数没有返回值（包括void类型）。

（3）析构函数没有任何参数。

（4）析构函数不能重载。

（5）一个类只能有一个析构函数。

与构造函数一样，如果类中没有定义析构函数，那么系统会自动定义一个析构函数，但这个析构函数是一个空函数，它不会做任何动作。

析构函数在类中的基本格式如下：

```
class TclassName {
  …
  public:
  …
  /* 析构函数 */
~ TclassName () ;
};


/* 析构函数的实现 */
TclassName:: TclassName ()
{ … }
```

实例 2-5   析构函数的使用。

```
#include <iostream.h>
using namespace std;


void useClass();

/* 定义类 */
class TStudentInfo{
    public:
        /* 构造函数原型 */
        TStudentInfo(int NO,float Chinese,float Math,float English);

        /* 析构函数原型 */
        ~TStudentInfo();

        /* 实现输出功能的函数 */
        void coutValue();
    private:
        int stNO;
        float stChinese;
        float stMath;
        float stEnglish;
};

/* 构造函数的实现 */
```

```
TStudentInfo::TStudentInfo(int NO,float Chinese,float Math,float English){
    stNO=NO;
    stChinese=Chinese;
    stMath=Math;
    stEnglish=English;
}
```

/* 实现析构函数 */
```
TStudentInfo::~TStudentInfo(){
    cout<<" 析构函数: "<<endl;
    cout<<" 此处可回收类中使用的内存 "<<endl;
}
```

/* 实现输出功能的函数 */
```
void TStudentInfo::coutValue(){
    cout<<"你输入的信息如下: "<<endl;
    cout<<" 学号: "<<stNO<<endl;

    cout<<" 语文成绩: "<<stChinese<<endl;
    cout<<" 数学成绩: "<<stMath<<endl;
    cout<<" 英语成绩: "<<stEnglish<<endl<<endl;
}
```

```
void useClass(){
    TStudentInfo st(10000,100,99,99.5);
    st.coutValue();
}
```

```
main(){
    cout<<"**********************************"<<endl;
    cout<<"*              析构函数的使用            *"<<endl;
    cout<<"**********************************"<<endl;
    useClass();
    cout<<"**********************************"<<endl;
}
```

程序运行结果如下:

```
**********************************
*          析构函数的使用           *
**********************************
你输入的信息如下：
学号：10000
语文成绩：100
数学成绩：99
英语成绩：99.5


析构函数：
此处可回收类中使用的内存

**********************************
```

可以看到，在调用类结束后系统将自动运行析构函数。

2.5 继承与派生

继承是面向对象中一个非常重要的概念，它描述了类与类之间的关系。当一个类拥有另一个类的所有数据和操作时，就称这两个类之间具有继承关系。其中，被继承的类称为父类（也称为超类、基类），继承的类称为子类。子类继承了父类的所有属性和方法，同时又具有自己的属性和方法。

利用类的继承机制，可以从已有的类中派生出新类。这种通过基类或父类继承而产生新类的过程称为派生。

前面已经讲过，类的成员可以通过访问限制符声明为public、private和protected 3种。它们对应的类继承也可以分为3种：

· 公用继承 基类的所有成员在派生类中的身份都不变，派生类可以引用基类的公用成员和保护成员，但不能引用基类的私有成员。

· 保护继承 派生类可以访问基类的公用成员和保护成员，但引用后基类的公用成员在派生类中变为保护成员，基类的保护成员在派生类中变为私有成员。

· 私有继承 派生类可以访问基类的公用成员和保护成员，但基类的公用成员和保护成员都成为派生类的私有成员。

派生类的定义格式如下：

```
class <派生类名>:<继承方式><基类名>{
    <派生类成员>
};
```

从一个类派生出另一个类的例子如下：

```
class TBaseClass{…};
```

```
…
/* 公用继承 */
class Class1:public TBaseClass {…} ;
…
/* 保护继承 */
class Class2:protected TBaseClass {…} ;
…
/* 私有继承 */
class Class3:private TBaseClass {…} ;
…
```

需要注意的是，作为基类，它必须有完整的类定义和实现过程，而不能仅有一个类名。

实例2-6　析构函数的使用。

```cpp
#include <iostream.h>
#include <string.h>

using namespace std;

/* 定义基类 */
class TStudent{
    public:
        /* 建立构造函数 */
        TStudent()  {
            /* 完成成员变量的赋值操作 */
            NO=1000;
            Chinese=100;
            Math=99;
            English=99.5;
            sum=Chinese+Math+English;
        }

        /* 建立析构函数 */
        ~TStudent()  {
            /* 将成员变量的值都置 0*/
            NO=0;
            Chinese=0;
            Math=0;
```

```
            English=0;
            sum=0;
        }
    protected:
        int NO;
        float Chinese,Math,English;
        float sum;
};
```

/* 定义派生类 */
```
class TStudent_output:public TStudent{
    public:
        void coutInfo();
};
```

/* 实现派生类的成员函数 */
```
void TStudent_output::coutInfo(){
    cout<<"学    号："<<NO<<endl;
    cout<<"=============="<<endl;
    cout<<"语文成绩："<<Chinese<<endl;
    cout<<"数学成绩："<<Math<<endl;
    cout<<"英语成绩："<<English<<endl;
    cout<<"总 成 绩："<<sum<<endl<<endl;
}
```

```
main(){
    cout<<"********************************"<<endl;
    cout<<"*           简单的派生类应用          *"<<endl;
    cout<<"********************************"<<endl;
    TStudent st;
    TStudent_output st1;
    st1.coutInfo();
    cout<<"********************************"<<endl;
}
```

程序运行结果如下：

```
********************************
```

```
*            简单的派生类应用           *
*****************************
学    号:1000
==============
语文成绩:100
数学成绩:99
英语成绩:99.5
总 成 绩:298.5


*****************************
```

在这个例子中,成员值的输出函数是放在TStudent类的派生类TStudent_output中来实现的。可以看到,在TStudent_output类中可以使用TStudent类中的公用成员和保护成员,请读者根据程序的运行结果自行分析程序。

类可以由一个基类派生,也可以由多个基类派生。由一个基类派生的类称为单继承类,而由多个基类派生的类称为多继承类。

多继承派生类的定义格式如下:

```
class <派生类名>:<继承方式><基类名1>,<继承方式><基类 2>…{
   <派生类成员>
}
```

2.6　友元

类的成员可以由访问限制符public、protected和private分为3类,其中被protected和private修饰的类成员只能被类的成员函数访问,它们具有隐蔽性。友元是C++提供给外部类或函数访问类的私有成员和保护成员的一种途径。

友元函数在类体中定义,但是它并不是类的成员函数,它只是被声明为类的友元函数的普通函数。友元函数使用修饰符friend来说明,函数名可以与类成员的函数同名,但它们是不同的函数。

友元函数在为外部函数提供访问类的私有成员和保护成员的同时,也带来了相应的弊端:友元函数的存在破坏了类的封装性。

友元函数的声明方式如下:

```
class TclassName{
  public:
    …
    /* 说明友元函数 */
    friend void friendFun();
```

```
        …
    private:
        …
};


/* 实现友元函数 */
void friend frendFun(){
    …

}
```

需要注意的是，由于友元函数不是类的成员函数，而只是被说明为友元函数的普通函数，所以在类体外实现这个函数的时候，不必像类的成员函数那样使用符号"::"，只需像普通函数一样实现即可。

实例2-7 友元函数的使用。

```
#include <iostream.h>
#include <string.h>

using namespace std;

class TStudent{
    public:
        TStudent()  {
            NO=10000;
            Chinese=100;
            Math=99;
            English=99.5;
        }
        ~TStudent()  {
            NO=0;
            Chinese=0;
            Math=0;
            English=0;
        }

        void showValue();
        friend void showValue();
    private:
```

```
        int NO;
        float Chinese,Math,English;
        float sum;
};

/* 实现类的成员函数 */
void TStudent::showValue(){
    sum=Chinese+Math+English;
}

/* 实现类的友元函数 */
void showValue(){
    TStudent st;
    st.showValue();

    cout<<"学      号："<<st.NO<<endl;
    cout<<"=========="<<endl;
    cout<<"语文成绩："<<st.Chinese<<endl;
    cout<<"数学成绩："<<st.Math<<endl;
    cout<<"英语成绩："<<st.English<<endl;
    cout<<"总 成 绩："<<st.sum<<endl;
}

main(){
    cout<<"********************************"<<endl;
    cout<<"*           友元函数的使用          *"<<endl;
    cout<<"********************************"<<endl;
    showValue();
    cout<<"********************************"<<endl;
}
```

程序运行结果如下：

```
********************************
*           友元函数的使用          *
********************************
学      号：10000

==========
```

语文成绩：100

数学成绩：99

英语成绩：99.5

总 成 绩：298.5

****

在这个例子的类体中定义了一个和类成员函数同名的友元函数，并且在类体外实现了这两个函数。通过程序的运行结果可以看到，类的友元函数成功地访问到了类中定义的私有成员变量。

2.7 上机操作

2.7.1 重载构造函数

函数重载的概念我们将在本书后面讲解，其基本概念是：两个函数具有相同的函数名，但是具有不同的参数表，例如：

```
fun1(int x,int y)
```

或者：

```
fun1(int x,int y,int z)
```

那么这两个函数就是重载函数，因为它们的参数列表是不同的。

实例2-8将演示类中构造函数的重载。

实例2-8 构造函数的重载。

```
#include <iostream.h>

#include <string.h>


class TStruFunc{
    public:
        /* 构造函数 */
        TStruFunc()  {
            a=0;
            b=0;
            sum=a+b;
        }


        /* 重载构造函数 */
        TStruFunc(int x, int y) {
            a=x;
            b=y;
```

```
      sum=a+b;
    }


    /* 输出成员值 */
    void coutFun()  {
      cout<<"a="<<a<<endl;
      cout<<"b="<<b<<endl;
      cout<<"sum="<<sum<<endl<<endl;
    }


  private:
    int a,b,sum;
};


main(){
    cout<<"*****************************"<<endl;
    cout<<"*          重载构造函数           *"<<endl;
    cout<<"*****************************"<<endl;
    TStruFunc sf1,sf2(100,200);
    cout<<"=== 调用第一个构造函数 ==="<<endl;
    sf1.coutFun();
    cout<<"=== 调用第二个构造函数 ==="<<endl;
    sf2.coutFun();
    cout<<"*****************************"<<endl;
}
```

程序运行结果如下：

```
*****************************
*          重载构造函数           *
*****************************
=== 调用第一个构造函数 ===
a=0
b=0
sum=0


=== 调用第二个构造函数 ===
a=100
```

```
b=200
sum=300


*****************************
```

2.7.2 利用析构函数判断数值大小

析构函数是系统在结束一个类的调用时自动调用的，在代码中不必也不允许再调用这个函数。实例2-9使用了析构函数来判断用户输入的数值大小。

实例2-9 利用析构函数判断数值大小。

```cpp
#include <iostream.h>
#include <string.h>

void inputFun();

class TStruFunc{
    public:
        /* 构造函数 */
        TStruFunc(int x, int y)  {
            a=x;
            b=y;
        }

        /* 输出成员值 */
        ~TStruFunc()  {
            if (a==b)
                cout<<"x=y";
            else if (a>b)
                cout<<"x>y";
            else
                cout<<"x<y";

            cout<<endl;
        }

    private:
        int a,b,sum;
```

```
};

void inputFun(){
    int x,y;
    cout<<"x=";
    cin>>x;
    cout<<"y=";
    cin>>y;

    TStruFunc(x,y);
}

main(){
    cout<<"*****************************"<<endl;
    cout<<"*          析构函数          *"<<endl;
    cout<<"*****************************"<<endl;
    inputFun();
    cout<<"*****************************"<<endl;
}
```

程序运行结果如下：

```
*****************************
*          析构函数          *
*****************************
x=100
y=90
x>y
*****************************
```

2.8　小结

本章讲述了面向对象的基本概念以及面向对象中的一些重要知识，例如对象、类、方法、继承等。深刻理解这些重要的概念是学习C++的基础，希望读者能够多动脑思考，多做实验。

本章带有大量的实例，通过分析这些实例可以更好更深刻地理解什么是对象，什么是类，类成员包括哪些内容，类成员是如何实现的等知识。

本章实例均可在Visual C++环境下实现。

2.9 习题

一、填空题

1. 在C++的成员函数中，有两种特殊的函数：_____函数和_____函数。

2. 构造函数与析构函数与其所在的类具有相同的_____。

3. 构造函数在该类的对象建立时被系统_____调用，不能被用户调用。因此，构造函数必须被声明为该类的_____。

4. 构造函数_____返回类型。

5. 构造函数可以带有参数，也可以没有参数，而析构函数则_____。

6. 析构函数名与类名_____，只是在类名前加一个_____符号，以区别于_____。

7. 一个类有_____个析构函数。

8. 构造函数_____重载，_____返回值。

9. 当一个类拥有另一个类的所有数据和操作时，就称这两个类之间具有_____关系。

10. 类成员的访问限制符可以分为：_____、_____和_____3种。

11. 友元函数就是在_____中定义的函数，但友元函数并不是类的_____函数，它只是被声明为类的友元函数的普通函数。友元函数使用修饰符_____来说明，

二、代码阅读题

分析下面的程序，写出程序运行的输出结果，并亲自上机验证。

代码1：

```cpp
#include <iostream.h>
#include <string.h>


float inputFun();


class TStruFunc{
    public:
        /* 构造函数 */
        TStruFunc()  {
          a=0;
          b=0;
        }


        /* 取最大值 */
        float max(float a, float b)  {
          float result;
```

```
        if (a>=b)
          result = a;
        else if (a<b)
          result = b;


        return result;
      }


      /* 析构函数 */
      ~TStruFunc()  {
        a=0;
        b=0;
      }

    private:
      int a,b;
};

float inputFun(){
    float x,y;

    cout<<"x=";
    cin>>x;
    cout<<"y=";
    cin>>y;

    TStruFunc sf;

    cout<<sf.max(x,y)<<endl;
}

main(){
    cout<<"*******************************"<<endl;
    cout<<"*          XXXXXXXX 的类        *"<<endl;
    cout<<"*******************************"<<endl;
    inputFun();
```

```
        cout<<"*****************************"<<endl;
    }
```

代码2：

```cpp
#include <iostream.h>
#include <string.h>

float inputFun();

class TStruFunc{
    public:
        /* 构造函数 */
        TStruFunc()  {
            length=0.0;
            width=0.0;
        }

        /* 求矩形面积 */
        float areaValue(float a, float b)  {
            length=a;
            width=b;
            area=a*b;

            return area;
        }

        /* 求周长 */
        float circValue(float a, float b)  {
            length=a;
            width=b;
            circ=2*(a+b);

            return circ;
        }

        /* 析构函数 */
        ~TStruFunc()  {
```

```
            length=0.0;
            width=0.0;
         }

     private:
        float length,width;
        float area,circ;
};


float inputFun(){
    float x,y;

    cout<<"请输入矩形长度：";
    cin>>x;
    cout<<"请输入矩形宽度：";
    cin>>y;

    TStruFunc sf;

    cout<<"矩形面积："<<sf.areaValue(x,y)<<endl;
    cout<<"矩形周长："<<sf.circValue(x,y)<<endl;
}


main(){
    cout<<"*******************************"<<endl;
    cout<<"*        求矩形面积和周长的类        *"<<endl;
    cout<<"*******************************"<<endl;
    inputFun();
    cout<<"*******************************"<<endl;
}
```

三、问答题

1．什么是类？类是一种数据类型吗？

2．类定义中的public、private和protected有什么区别？

3．面向对象概念中的继承的含义是什么？有几种继承方式？

第3章 C++程序的基本控制结构

本章学习目标

- ↳ 顺序结构
- ↳ 选择结构
- ↳ 循环结构
- ↳ switch语句
- ↳ 循环嵌套

通过前面两章的学习，已经基本了解了C++程序的程序结构、基本数据类型、面向对象编程思想和C++面向对象的编程方法。从本章开始，将学习C++编程的具体知识，首先学习C++程序的基本控制结构。

前面两章的主要内容是理论知识，从本章开始将学习具体的编程知识，需要指出的是，这些具体知识都是以编程思想为指导的，脱离了理论知识的指导，则无从谈起具体的技术手段，因此希望读者能够先深刻理解前两章的理论知识。

提示

所有程序语句都必须在一定的控制结构内执行，所谓程序的控制结构，简单地讲就是程序语句的执行顺序，程序语句按照什么顺序去执行是由程序的控制结构决定的。

3.1 程序结构知识

程序语句是按照一定顺序执行的，最简单的当然就是按照程序书写的顺序执行，这就是顺序结构。为了完成一定的目标，有些程序段需要反复执行，直到目标达成为止，这就是循环结构。

当一段程序在执行时，需要判断是否符合一定的条件，如果符合条件则执行一个程序段，相反则执行另外一个程序段，这就是选择程序结构。本节将学习C++的几种基本程序结构：顺序结构、选择结构和循环结构。

3.2 顺序结构

顺序结构是所有程序控制结构中最为简单的一种结构。顺序结构就是程序按照程序语句的书写顺序来执行。

顺序结构的流程图如图3-1所示。

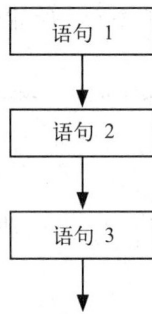

图3-1　顺序结构流程图

下面是一个顺序控制结构的例子。

实例3-1　C++的顺序控制结构。

```cpp
#include <iostream.h>

short int a,b;

main(){
    short int c,d;
    a = 0;
    b = 1;
    c = 2;
    d = a + b + c;
    cout <<"a ="<<a<<"\n";
    cout <<"b ="<<b<<"\n";
    cout <<"c ="<<c<<"\n";
    cout <<"d = a + b + c ="<<d<<"\n";
    return 0;
}
```

程序运行结果如下：

```
a = 0
b = 1
c = 2
d = a + b + c = 3
```

这个例子非常容易理解，程序首先将头文件iostream.h包含进来，这个头文件包含了C++基本的输入/输出操作。第二步程序定义了两个short int类型的全局变量a和b，这两个变量将在后面的程序语句中使用。接下来进入程序的主体，也就是main主函数段。在主函数中，程序首先定义了两个short int类型的局部变量c和d。接下来的三条语句为全局变量a和b以及局部变量c赋值。

而后将全局变量a和b以及局部变量c相加，并将结果赋值给局部变量d。接下来的四条语句将变量a、b、c和d的值输出到屏幕上，以供用户查看，然后给出返回值0，表示执行完成。

3.3 选择结构

C++的选择结构与C类似，分条件选择结构（if … else）和switch结构。

3.3.1 if结构

if语句根据程序给出的条件决定要执行的操作。if语句有3种形式，下面分别做介绍。

1. If (条件表达式) 语句

这是最简单的if语句，它首先判断条件表达式是否成立，如果成立则执行后面跟着的语句，例如：

```
if (a>b)  c=a;
```

需要指出的是，如果条件表达式成立之后，后面需要执行的是一个程序段，则这个程序段需要用一对花括号括起来，例如：

```
if (a>b)  {
    c=a;
    cout <<"max ="<<c<<"\n";
}
```

if语句的程序结构流程图如图3-2所示。

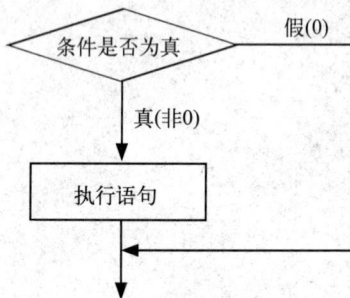

图3-2　if语句的程序结构流程图

2. If (条件表达式) 语句1 else 语句2

这种结构的if语句也是首先判断条件表达式是否成立，如果条件成立则执行语句1，否则执行语句2。需要说明的是，else只能和if语句搭配使用，如果语句1和语句2代表的是程序段，那么也必须用花括号把程序段括起来，否则C++会认为只执行语句段中的第一条语句。

if…else语句的程序结构流程图如图3-3所示。

图3-3　if…else语句的程序结构流程图

3. 多重选择结构

这种if语句有多个条件判断，例如：

```
if（条件表达式1）语句1
else if（条件表达式2）语句2
else if（条件表达式3）语句3
  …
else if（条件表达式n）语句n
```

这种if语句结构和第二种if语句十分相似，只不过在else后又加了if语句，再判断新的条件表达式，以此类推。图3-4是这种程序结构的流程图。

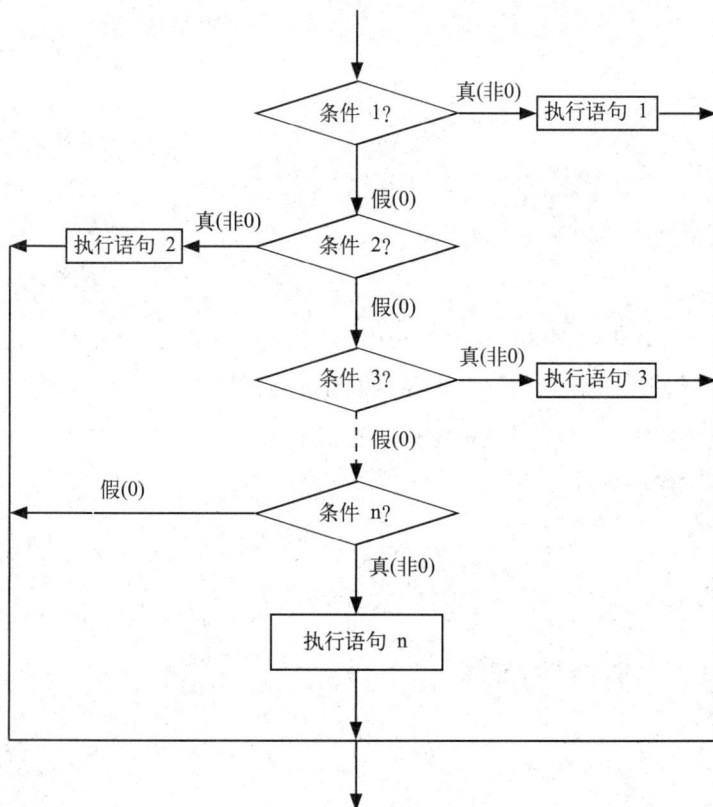

图3-4　多重条件的if语句程序结构流程图

下面3个例子分别为这3种if结构的运用。

实例3-2 根据输入数据输出其中较大的值，方法1。

```
#include <iostream.h>


main(){
    short int a,b;
    cout<<"***************"<<"\n";
    cout<<"a =";
    cin>>a;
    cout<<"b =";
    cin>>b;
    if (a>b)
        cout<<"max ="<<a<<"\n";
    if (a<b)
        cout<<"max ="<<b<<"\n";
    cout<<"***************"<<"\n";
    return 0;
}
```

程序运行结果如下：

```
***************
a = 2
b = 3
max = 3
***************
```

这个例子中的if语句用来判断用户输入的值中哪个较大，并且将较大的那个值输出到屏幕上。
下面将这个例子修改一下，用if … else结构同样能达到目的。

实例3-3 根据输入数据输出其中较大的值，方法2。

```
#include <iostream.h>


main(){
    short int a,b;
    cout<<"***************"<<"\n";
    cout<<"a =";
    cin>>a;
    cout<<"b =";
    cin>>b;
```

```
    if (a>b)
      cout<<"max ="<<a<<"\n";
    else
      cout<<"max ="<<b<<"\n";
    cout<<"***************"<<"\n";
    return 0;
}
```

程序运行结果如下：

```
***************
a = 2
b = 3
max = 3
***************
```

通过结果可以看到，两个用不同if结构编写的程序所实现的结果是一样的。下面再将这个例子修改一下，同样能够得到这个结果，而且能够判断出用户所输入的数值是不是等值的。

实例3-4　根据输入数据输出其中较大的值，方法3。

```
#include <iostream.h>

main(){
    short int a,b;
    cout<<"***************"<<"\n";
    cout<<"a =";
    cin>>a;
    cout<<"b =";
    cin>>b;
    if (a>b)
      cout<<"max ="<<a<<"\n";
    else if (a<b)
      cout<<"max ="<<b<<"\n";
    else if( a==b)
      cout<<"a=b"<<"\n";
    cout<<"***************"<<"\n";
    return 0;
}
```

程序运行结果如下：

```
***************
```

```
a = 3

b = 3

a=b

***************
```

这个例子使用了第三种if语句结构，在实例3-2和实例3-3的基础上，还实现了判断用户输入的两个值是否等值的功能。

下面这个例子略显复杂，它综合使用了if语句的3种结构。

实例3-5 判断输入的年份是不是闰年。

```
#include <iostream.h>

main(){
    short int year,a;
    cout<<"***********************************"<<"\n";
    cout<<" 请输入你要判断的年份：";
    cin>>year;
    if (year%4==0)  {
      if (year%100==0)  {
        if (year%400==0)
           a=1;
         else
           a=0;
       }
      else
        a=1;
    }
    else
      a=0;

    if(a==1)
      cout<<" 你输入的年份："<<year<<" 是闰年。"<<"\n";
    else
      cout<<" 你输入的年份："<<year<<" 不是闰年。"<<"\n";
    cout<<"***********************************"<<"\n";
}
```

程序运行结果如下：

```
***********************************
```

```
请输入你要判断的年份:2020
你输入的年份:2020 是闰年。
*****************************
```

在数学概念中，闰年的规则是：一般年份凡能被4整除的都是闰年，但整百的年必须能被400整除才是闰年。

实例3-5利用if语句的嵌套，并根据数学概念中判断闰年的规则来判断用户输入的年份是不是闰年。

3.3.2　switch结构

实例3-5中使用了if语句的嵌套来解决多重条件判断的问题，应该说这是一种解决问题的好方法，但是并不是最好的解决方法，因为这样做会使代码比较冗长，特别是当需要判断的条件的取值比较多的时候。

C++提供了switch语句来处理这种问题，使用switch可以使程序看起来更简洁，更容易让人理解。

switch语句的一般形式如下：

```
switch（条件表达式）{
   case 常量表达式 1: 语句1;
   case 常量表达式 2: 语句2;
   …
   常量表达式 n: 语句n;
   default: 语句n+1;
}
```

下面这个例子展示了switch语句的基本用法。

实例3-6　根据用户输入的字符，用switch语句判断要执行的操作。

```cpp
#include <iostream.h>

void funSwitch();

void funSwitch(){
    short int a,b;
    char c;
    a = 8;
    b = 9;
    cout<<"*****************************"<<"\n";
    cout<<"+"<<"\n";
    cout<<"-"<<"\n";
```

```
        cout<<"*"<<"\n";
        cout<<"/"<<"\n";
        cout<<" 请输入你要执行的计算：";
        cin>>c;

        switch (c)  {
            case'+': cout<<" 你选择了加法操作："<<a<<"+"<<b<<" ="<<a+b<<"\n";
            case'-': cout<<" 你选择了减法操作："<<a<<"-"<<b<<" = "<<a-b<<"\n";
            case'*': cout<<" 你选择了乘法操作："<<a<<"*"<<b<<" = "<<a*b<<"\n";
            case'/': cout<<" 你选择了除法操作："<<a<<"/"<<b<<" = "<<a/b<<"\n";
            default: cout<<" 你输入的字符不在选择范围内！ "<<"\n";
        }

        cout<<"***************************"<<"\n";
}


main(){
    funSwitch();
    return 0;
}
```

实例3-6的目的是，根据用户输入的字符来进行相应的操作。其执行结果如下：

```
***************************
+
-
*
/
请输入你要执行的计算：+
你选择了加法操作：8+9 = 17
你选择了减法操作：8-9 = -1
你选择了乘法操作：8*9 = 72
你选择了除法操作：8/9 = 0
你输入的字符不在选择范围内！
***************************
```

从实例3-6的执行结果看，用户输入的字符+代表要执行加法操作，但是程序把所有的操作（包括default后面的语句）都执行了，为什么？再执行一遍，这次输入字符-：

```
***************************
```

```
+
-
*
/
```

请输入你要执行的计算：-

你选择了减法操作：8-9 = -1

你选择了乘法操作：8*9 = 72

你选择了除法操作：8/9 = 0

你输入的字符不在选择范围内！

```
*************************
```

这次执行结果是，除了加法操作之外所有的语句都被执行了，为什么？其原因在于：case
常量表达式仅仅起到语句标号的作用，并不在该处进行判断，执行switch语句时，根据switch后
面表达式的值找到匹配的入口标号，便从此标号开始执行下去，不再进行判断。因此需要在执
行完一个case分支后，使流程跳出switch结构，终止switch语句的执行。C++提供了break语句来
达到此目的，可将switch语句的一般形式修改为：

```
switch（条件表达式）{
  case 常量表达式1：语句1；break;
  case 常量表达式2：语句2；break;
  …
  常量表达式n：语句n；break;
  Default：语句n+1；
}
```

下面将实例3-6修改一下，以达到选择操作的目的。

实例3-7　根据用户输入的字符，用switch和break语句判断要执行的操作。

```
#include <iostream.h>

void funSwitch();

void funSwitch(){
    short int a,b;
    char c;
    a = 8;
    b = 9;
    cout<<"*************************"<<"\n";
    cout<<"+"<<"\n";
    cout<<"-"<<"\n";
```

```
    cout<<"*"<<"\n";
    cout<<"/"<<"\n";
    cout<<" 请输入你要执行的计算：";
    cin>>c;

    switch (c)  {
        case'+': cout<<"你选择了加法操作："<<a<<"+"<<b<<" ="<<a+b<<"\n";break;
        case'-': cout<<"你选择了减法操作："<<a<<"-"<<b<<" = "<<a-b<<"\n";break;
        case'*': cout<<"你选择了乘法操作："<<a<<"*"<<b<<" = "<<a*b<<"\n";break;
        case'/': cout<<"你选择了除法操作："<<a<<"/"<<b<<" = "<<a/b<<"\n";break;
        default: cout<<"你输入的字符不在选择范围内！"<<"\n";
    }

    cout<<"****************************"<<"\n";
}


main(){
    funSwitch();
    return 0;
}
```

这次的程序执行结果如下：

```
****************************
+
-
*
/
请输入你要执行的计算：+
你选择了加法操作：8+9 = 17
****************************
```

如果输入的字符不在程序提示的字符+、-、*、/中，则程序执行结果如下：

```
****************************
+
-
*
/
请输入你要执行的计算：a
```

你输入的字符不在选择范围内!

****

　　实例3-7其实是用C++编写的一个简化的计算器程序，但是这里有一个问题：要计算的两个数值是用户输入的，如果涉及除法运算和求余运算，则除数不能为0，这是数学规定。因此，可将实例3-7再修改一下，使其更符合数学规则和实际情况。

　　实例3-8　简单的计算器程序。

```
#include <iostream.h>

void funSwitch();

void funSwitch(){
    short int a,b;
    char c;

    cout<<"****************************"<<"\n";
    cout<<" 请输入变量 a 的值：";
    cin>>a;
    cout<<" 请输入变量 b 的值：";
    cin>>b;
    cout<<"\n";
    cout<<"+"<<"\n";
    cout<<"-"<<"\n";
    cout<<"*"<<"\n";
    cout<<"/"<<"\n";
    cout<<"%"<<"\n";
    cout<<" 请输入你要执行的计算：";
    cin>>c;

    switch (c)  {
        case'+': cout<<" 你选择了加法操作："<<a<<"+"<<b<<" = "<<a+b<<"\n";break;
        case'-': cout<<" 你选择了减法操作："<<a<<"-"<<b<<" = "<<a-b<<"\n";break;
        case'*': cout<<" 你选择了乘法操作："<<a<<"*"<<b<<" = "<<a*b<<"\n";break;
        case'/':
            if (b==0)
                cout<<" 除数不能为 0!"<<"\n";
            else
```

```
                    cout<<" 你选择了除法操作："<<a<<"/"<<b<<" = "<<a/b<<"\n";
                    break;
            case'%':
                if (b==0)
                    cout<<" 除数不能为 0！ "<<"\n";
                else
                    cout<<" 你选择了求余操作："<<a<<"%"<<b<<" = "<<a%b<<"\n";
                    break;
            default: cout<<" 你输入的字符不在选择范围内！ "<<"\n";
        }


        cout<<"***************************"<<"\n";
}


main(){
    funSwitch();
    return 0;
}
```

实例3-8在实例3-7的基础上做了修改，主要是增加了求余操作和在除法运算以及求余操作的时候能够先对除数是否为0的判断，这样就能够避免做除法操作和求余操作时出错，使程序符合数学规则和实际应用。

> **提示**
>
> 前面讲述的if选择结构是C++的基本选择结构，它可以代替switch结构，但是从程序的简洁性、易读性来讲，仍应该在条件的多重判断时使用switch语句，这样的程序更利于维护。

3.4 条件运算符?:

前面已经学习了if…else结构，它可以实现条件判断。另外，C++还提供了一种更加简洁的条件运算符?:，同样能够达到此目的。

条件运算符的一般形式如下：

```
exp1?exp2:exp3
```

可以看到，条件运算符是一个三目运算符。在这个条件表达式中，如果表达式exp1成立（即值为真），则执行exp2，整个条件表达式的值为exp2的值；否则执行exp3，整个条件表达式的值为exp3的值。

实例3-3演示了if…else的用法，下面把这个例子用条件运算符来重写。

实例3-9　用条件运算符判断两个数值的大小。

```
#include <iostream.h>


main(){
    short int a,b;
    cout<<"***************"<<"\n";
    cout<<"a =";
    cin>>a;
    cout<<"b =";
    cin>>b;

    (a>b)?cout<<"max ="<<a<<"\n":cout<<"max ="<<b<<"\n";
    cout<<"***************"<<"\n";
    return 0;
}
```

程序运行结果如下：

```
***************
a = 1
b = 2
max = 2
***************
```

C++规定，条件运算符的运算顺序是从左到右，即先运算?左边的表达式，然后根据左边表达式的运算结果（真或假）来决定执行?右边两个表达式中的哪一个。

在这个例子中，(a>b)?cout<<"max = "<<a<<"\n":cout<<"max = "<<b<<"\n";语句代表了实例3-3中if…else的功能。程序先运算(a>b)，如果为真，则执行cout<<"max = "<<a<<"\n，否则执行cout<<"max = "<<b<<"\n"。

在使用条件运算符时，应注意以下几点：

（1）条件运算符的优先级低于关系运算符和算术运算符，但高于赋值运算符。

（2）条件运算符的?号和:号是成对出现的，不能分开单独使用。

（3）条件运算符的结合方向是自右至左。

在上例中，我们完全可以不要?左边表达式的括号，将(a>b)变为a>b，这样计算结果是一样的，但这样却并不能使程序的结构显得更清晰、更容易被理解。C++规定了运算符的运算顺序，但在实际应用中往往以圆括号来区分运算符的执行顺序。

下面这个例子演示了条件运算符的优先级。

实例3-10　条件运算符的优先级。

```
#include <iostream.h>
```

```
main(){
    int a, b, c, d, max1,max2;
    cout<<"***********************"<<"\n";
    cout<<" 请为变量 a 赋值  a = ";
    cin>>a;
    cout<<" 请为变量 b 赋值  b = ";
    cin>>b;
    cout<<" 请为变量 c 赋值  c = ";
    cin>>c;
    cout<<" 请为变量 d 赋值  d = ";
    cin>>d;

    max1=a>b?a:(c>d?c:d);
    max2=a>b?a:c>d?c:d;
    cout<<"max1="<<max1<<"\n";
    cout<<"max2="<<max2<<"\n";
    cout<<"***********************"<<"\n";
}
```

程序运行结果如下：

```
***********************
请为变量 a 赋值  a = 9
请为变量 b 赋值  b = 8
请为变量 c 赋值  c = 7
请为变量 d 赋值  d = 6
max1=9
max2=9
***********************
```

在实例3-10中定义了两个变量max1和max2，用来存放条件表达式的计算结果。虽然这两个条件表达式的计算都是一样的，但从程序的可读性上来看，显然第二个更容易理解。

3.5 循环结构

C++提供的循环结构与C类似，包括for循环、while循环以及do while循环等结构。

3.5.1 while语句

while语句是C++中的循环语句，其一般形式为：

```
while（条件表达式）语句；
```

其中，条件表达式是要执行while循环所必须满足的条件，当条件满足时，执行后面的语句或语句组，否则不执行。这样的语句或语句组称为循环体。

while语句的流程图如图3-5所示。

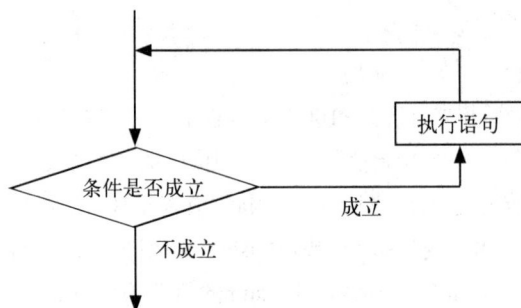

图3-5　while循环语句的流程图

通过下面这个例子来说明while循环。

实例3-11　用while循环求100以内所有的偶数和。

```cpp
#include <iostream.h>

void myFun();

void myFun() {
    short int a,sum;

    a = 1;
    sum = 0;

    cout<<"***********************"<<"\n";
    while (a<=100)  {
      if (a%2==0)
        sum = sum + a;
        a++;
      }
    cout<<"sum ="<<sum<<"\n";
    cout<<"***********************"<<"\n";
}

main(){
    myFun();
```

```
    return 0;
}
```

程序运行结果如下：

```
************************

  sum = 5050

************************
```

这个例子使用了while循环来计算1～100之间的偶数之和。首先定义两个short int类型的变量a和sum，其中a是循环变量，它的初始值是1，并且由a++语句控制变化。每完成一次循环后其值自动加1，使循环变量趋向于它的最终值100。当a的值等于101的时候，循环变量已经不能满足<=100的条件，则退出循环体。变量sum是普通的short int型变量，它的作用是在每次循环后将自身的值加上a的值。这样，当while循环结束时，sum的值即为1～100之间所有偶数的和。

使用while语句时应注意以下两点：

（1）while循环必须有一个循环变量，而且必须确定循环变量的初始值。

（2）每次执行一遍循环体后，循环控制变量要变化一次，其变化的趋势是使自身的值趋向于最终值。

下面再看一个稍微复杂的例子。

实例3-12　用while语句输出九九乘法表。

```
#include <iostream.h>


void myFun();


void myFun(){
    int a = 1;
    int b = 1;
    int sum = 0;
    cout<<"*************************************************************"<<"\n";
    cout<<"*                        九九乘法表                        *"<<"\n";
    cout<<"*************************************************************"<<"\n";

    while (a<=9)  {
        b=1;
        while (b<=a)  {
            sum=a*b;
            cout<<b<<"*"<<a<<"="<<sum<<"  " ;
            if (b==a)
                cout<<"\n";
```

```
            b++;
        }
        a++;
    }

    cout<<"*************************************************************"<<"\n";
}

main(){
    myFun();
    return 0;
}
```

程序运行结果如下：

```
*************************************************************
*                     九九乘法表                            *
*************************************************************
1*1=1
1*2=2 2*2=4
1*3=3 2*3=6 3*3=9
1*4=4 2*4=8 3*4=12 4*4=16
1*5=5 2*5=10 3*5=15 4*5=20 5*5=25
1*6=6 2*6=12 3*6=18 4*6=24 5*6=30 6*6=36
1*7=7 2*7=14 3*7=21 4*7=28 5*7=35 6*7=42 7*7=49
1*8=8 2*8=16 3*8=24 4*8=32 5*8=40 6*8=48 7*8=56 8*8=64
1*9=9 2*9=18 3*9=27 4*9=36 5*9=45 6*9=54 7*9=63 8*9=72 9*9=81

*************************************************************
```

实例3-12使用了两个while循环，分别控制变量a和b的值，外层循环控制变量a的值，使其值在从1逐步自加1直到9。内层循环控制变量b的值，使其值不能大于变量a的值。内层循环同时完成了九九乘法表计算的功能。

3.5.2　do…while语句

C++循环结构的另外一种结构是do…while循环，这种循环和while循环不同的地方在于：while循环首先判断循环条件是否成立，如果不成立则不执行循环体。而do…while循环则不同，它首先执行一遍循环体，然后再判断循环条件是否成立，如果成立则继续执行循环体，直到循环条件不能满足为止。也就是说，无论循环条件成立与否，do…while的循环体都至少要被执行一次。

do…while循环语句的一般形式如下：

```
do {
   语句；
} while （条件表达式）
```

do…while循环的流程图如图3-6所示。

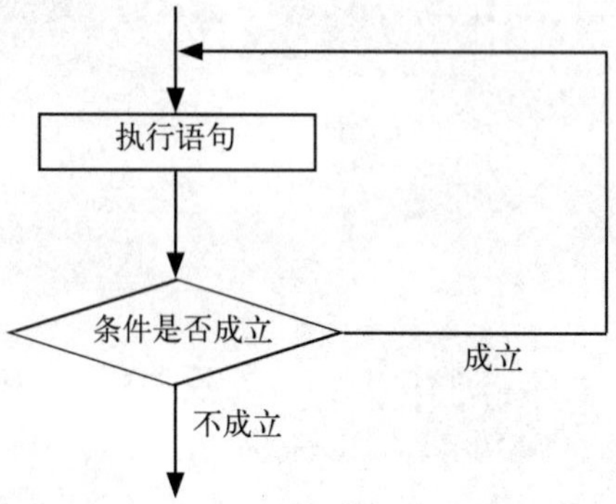

图3-6　do…while循环流程图

下面的例子是用 do…while 循环来求 1 ~ 100 之间的偶数和。

实例 3-13　用 do…while 循环求 1 ~ 100 之间的偶数和。

```cpp
#include <iostream.h>

void myFun();

void myFun() {
    short int a,sum;

    a = 1;
    sum = 0;

    cout<<"************************"<<"\n";
    do  {
      if (a%2==0)
        sum = sum + a;
        a++;
    }while (a<=100);

    cout<<"sum ="<<sum<<"\n";
    cout<<"************************"<<"\n";
}
```

```
main(){
    myFun();
    return 0;
}
```

程序运算结果如下：

```
************************
 sum = 5050
************************
```

实例3-13的循环首先执行一遍循环体，a的值自动加1，然后再判断是否满足循环条件。

下面的两个例子演示了while和do…while循环的不同点。

实例3-14　用while循环求指定数值段的值。

```
#include <iostream.h>

void myFun();

void myFun(){
    short int a,sum;
    sum = 0;
    cout<<"************************"<<"\n";
    cout<<"请为变量a赋值 a = ";
    cin>>a;
    while (a<=5)  {
        sum = sum + a;
        a++;
    }

    cout<<"sum ="<<sum<<"\n";
    cout<<"************************"<<"\n";
}

main(){
    myFun();
    return 0;
}
```

实例3-14运行时需要用户先为变量a赋值，同时变量a又是while循环的循环变量，while循环的循环体是将a的值累加到变量sum中。分别为变量a赋予小于5和大于5的两个值，其运行结果分别如下：

```
*********************
请为变量a赋值 a = 3
sum = 12
*********************


*********************
请为变量a赋值 a = 6
sum = 0
*********************
```

实例3-15　用 do…while 循环求指定数值段的值。

```
#include <iostream.h>

void myFun();

void myFun(){
    short int a,sum;
    sum = 0;
    cout<<"*********************"<<"\n";
    cout<<" 请为变量a赋值 a = ";
    cin>>a;
    do  {
        sum = sum + a;
        a++;
    }while (a<=5);

    cout<<"sum ="<<sum<<"\n";
    cout<<"*********************"<<"\n";
}


main(){
    myFun();
    return 0;
}
```

运行实例3-15，分别为变量a赋予小于5和大于5的两个值，其运行结果分别如下：

```
*********************
请为变量a赋值 a = 3
sum = 12
*********************


*********************
请为变量a赋值 a = 6
sum = 6
*********************
```

比较实例3-14和实例3-15可以看到，当为变量a赋予的值小于使循环条件成立的最大值5时，这两个程序的执行结果是一样的。当为变量a赋予的值大于使循环条件成立的最大值5时，则这两个程序的执行结果是不一样的。

这是因为：在while循环中，程序会首先判断循环条件是否成立，因输入的值为6时大于最大值5，所以while的循环条件不成立，因此程序不执行while的循环体，sum的值仍为初始值0。而在do…while循环中，程序首先执行一遍循环体，再判断循环条件是否成立，尽管输入的值6大于最大值5，但是do…while的循环体依然要被执行一次，所以sum的值为sum的初始值0加上a的初始值6，结果为6。

下面是一个比较复杂的do…while循环的例子。

实例3-16　红星化工厂污水系数统计程序。

```cpp
#include <iostream.h>

void myFun();

void myFun(){
    float sum,score;
    short int num;
    short int i;

    char yN;

    do {
        sum = 0.0;
        i = 1;

        cout<<" 请输入污水指标统计系列 (1~5):";
        cin >> num;
```

```
        cout <<" 你选择的是统计污水指标系列 "<<num<<"\n";
        while ( i <= num )  {
            cout <<" 请输入第 " << i <<"个参数 :";
            cin >> score;
            sum += score;
            i++;
        }
        cout <<" 水指标统计系列 "<<num<<" 的指标总额为 :"<< sum <<"\n";
        cout <<" 你还需要统计别的系列吗 (Y/N)?";
        cin  >> yN;
    }while( yN =='y' || yN == 'Y');
}


main(){
    cout<<"**********************************************"<<"\n";
    cout<<"*           红星化工厂污水指标统计           *"<<"\n";
    cout<<"**********************************************"<<"\n";
    myFun();
    cout<<"**********************************************"<<"\n";
    return 0;
}
```

在这个例子中，do…while循环内部还有一个while循环。while循环的作用是要求用户逐个输入相应的参数。在一个while循环完成后做出对相关参数的累加计算，在这个过程完成后，由do…while循环根据用户输入来决定是否再执行一遍do…while循环，以使用户能够连续使用本程序。

程序运行结果如下：

```
**********************************************
*           红星化工厂污水指标统计           *
**********************************************
请输入污水指标统计系列 (1~5) : 5
你选择的是统计污水指标系列 5
请输入第 1 个参数 : 2.5
请输入第 2 个参数 : 3.1
请输入第 3 个参数 : 1.9
请输入第 4 个参数 : 2.0
请输入第 5 个参数 : 1.7
```

水指标统计系列 5 的指标总额为：11.2

你还需要统计别的系列吗（Y/N)?y

请输入污水指标统计系列（1~5)：2

你选择的是统计污水指标系列2

请输入第1个参数：1.1

请输入第2个参数：3.1

水指标统计系列 2 的指标总额为：4.2

你还需要统计别的系列吗（Y/N)?N

****

3.5.3　for语句

C++循环结构还包括for循环。for循环是C++循环结构中最灵活、使用最频繁的循环语句。for循环特别适用于已知循环次数的情况。

for循环的一般形式如下：

for（表达式1；表达式2；表达式3) 语句；

for循环的执行过程如下：

（1）先求解表达式1。

（2）求解表达式2，若其值为真（非0），则执行for语句中指定的内嵌语句，然后执行第（3）步；若其值为假（0），则结束循环，转到第（5）步。

（3）求解表达式3。

（4）转回第（2）步继续执行。

（5）循环结束，执行for语句之外的语句。

for语句最简单的应用形式也是最容易理解的形式如下：

for（循环变量赋初值；循环条件；循环变量增量） 语句；

循环变量赋初值是一个赋值语句，用来给循环控制变量赋初值；循环条件是一个关系表达式，它决定什么时候退出循环；循环变量增量，它定义了循环控制变量每循环一次后按什么方式变化。这3部分之间用分号分隔。

例如：

for (a=1; a<=5; a++) sum = sum+a;

先给a赋初值1，判断a是否小于或等于5，若是则执行语句，之后a的值增加1。再重新判断，直到条件为假，即a>5时,结束循环。

相当于：

```
a=1;
while (a<=5)  {
    sum=sum+a;
    a++;
}
```

下面来看一个简单的for循环的例子。

实例3-17 用for循环实现数字输出。

```
#include <iostream.h>


void myFun();


void myFun(){
    short int a;
    cout<<"********************************"<<"\n";
    for (a=1;a<=10;a++)   {
        cout<<"a ="<<a;
        if (a < 10)
            cout<<",";
        if (a==5)
            cout<<"\n";
    }

    cout<<"\n";
    cout<<"********************************"<<"\n";
}


main(){
    myFun();
    return 0;
}
```

程序运行结果如下：

```
********************************
a = 1,a = 2,a = 3,a = 4,a = 5,
a = 6,a = 7,a = 8,a = 9,a = 10
********************************
```

实例3-17是一个非常简单的for循环的例子，它的功能是按顺序输出1～10这10个自然数。在程序中，首先定义了一个short int类型的变量a，但是并没有给这个变量赋初值，而是在for循环的循环条件中为它赋初值。接下来在for (a=1; a<=10; a++)语句中完成了如下的任务：

（1）为循环变量a赋初值1。

（2）设定循环变量a的最大取值为10。

（3）使循环变量a自增1。

在循环体中，完成了依次输出1~10这10个自然数，并通过if语句来实现程序按格式输出。

使用for循环时需要注意以下几点：

（1）for循环的一般形式为for (表达式1; 表达式2; 表达式3) 语句。其中表达式1、表达式2和表达式3都是可选项，都可以省略，但是分号不能省略。

（2）通常，表达式1为循环变量赋初值，表达式2为循环条件，表达式3为循环变量的自增量。

（3）省略表达式1，表示不为循环变量赋初值。

```
for (; a<=10; a++) ;
```

但是循环变量必须在此语句前通过其他形式赋予初值，比如在循环变量定义时同时赋初值，或者在循环变量定义完成后，通过其他赋值语句为其赋初值。另外，循环变量的定义也可以在表达式1中完成。在实例3-17中，循环变量a的定义和赋初值也可以合成为一步由表达式1来完成。例如：

```
for (short int a=1; a<=10; a++)
```

这条语句等效于：

```
    short int a;
  for (a=1; a<=10; a++) ;
```

（4）省略表达式2时，表示必须在程序体中设置能使这个for循环跳出循环体的条件，否则for循环就会成为一个死循环，无止境地执行下去。

```
for (a=1; ; a++) ;
```

使用这种方式时可以在循环体中加入if语句来跳出for循环，这个问题将在后面章节中讲到。

（5）省略表达式3时，表示不对循环变量的值做更改。

```
for (a=1; a<10;)
```

这时可以在循环体中加入对循环变量做出更改的语句，否则这个for循环也会因为循环变量总能满足循环条件而成为死循环，例如：

```
for (a=1; a<=10;) {
  …
  a++;
  …
}
```

实例3-18　用 for 循环求 1 ~ 100 之间的偶数和。

```
#include <iostream.h>

void myFun();

void myFun() {
    short int a,sum;
```

```
        sum = 0;

        cout<<"************************"<<"\n";
        for (a==1; a<=100;)  {
            if (a%2==0)
                sum = sum + a;
            a++;
        }

        cout<<"sum ="<<sum<<"\n";
        cout<<"************************"<<"\n";
    }


main(){
    myFun();
    return 0;
}
```

程序运行结果如下：

```
************************
 sum = 5050
************************
```

本例比较简单，只是在for循环中将第三个表达式省略了，改为在循环体中让循环变量自加1。

⚙ 实例3-19   使用for循环实现的九九乘法表。

```
#include <iostream.h>


void myFun();

void myFun(){
    int sum = 0;
    cout<<"*****************************************************************"<<"\n";
    cout<<"*                    使用 for 循环实现九九乘法表                 *"<<"\n";
    cout<<"*****************************************************************"<<"\n";
    for (short int a=1; a<=9; a++)  `{
        for (short int b=1; b<=a; b++)  {
```

```
                sum=a*b;
                cout<<b<<"*"<<a<<"="<<sum<<"";
                if (b==a)
                    cout<<"\n";
            }
        }
        cout<<"*********************************************************"<<"\n";
}


main(){
    myFun();
    return 0;
}
```

本例使用for语句来实现前面用while语句实现的九九乘法表。在本程序中，把循环变量的定义放在for循环的表达式1中进行，而且本例中用到了两个for循环，一个for循环位于另外一个for循环的内部，这叫作循环嵌套。这种循环方式以后会经常用到，将会在后续章节学习中不断熟悉它。

3.6　循环嵌套

在实例3-19中使用了两个for循环，并且把一个for循环放在另一个for循环的内部，这种结构称为循环嵌套。循环嵌套的定义为：一个循环体内又包含另外一个完整的循环结构。如果内嵌的循环再嵌套其他循环则称为多层嵌套。在具体程序中，经常会用到循环嵌套。

实例3-20　求1!+2!+3!+…n!的值。

```
#include <iostream.h>


void myFun();


void myFun(){
    int i,j,n;
    float k,sum;


    cout<<" 请输入一个正整数值：";
    cin>>n;


    sum = 0;
```

```
    for (i=1;i<=n;i++)  {
        k = 1;
        for (j=1;j<=i;j++)
            k = k * j;
        sum = sum + k;
    }


    cout<<"1 到 "<<n<<" 的阶乘之和为：" <<sum<<"\n";
}


main(){
    cout<<"***************************"<<"\n";
    cout<<"*    求 1!+2!+3!+...+n! 的值    *" <<"\n";
    cout<<"***************************"<<"\n";
    myFun();
    cout<<"***************************"<<"\n";
}
```

程序运行结果如下：

```
***************************
*    求 1!+2!+3!+...+n! 的值     *
***************************
请输入一个正整数值：9
1 到 9 的阶乘之和为：409113
***************************
```

求阶乘的数学公式如下：

```
n! = 1*2*3*n
```

在本例中，首先定义了3个int型变量i、j和n。变量i和j是两个for循环的循环变量，变量n是用户要输入的值，表示要求到达的那个最大的数。接下来又定义了两个float型变量k和sum。变量k用来存放每个数字的阶乘数，变量sum用来存放每个数字阶乘的和，即最终需要得到的结果。

本例中的for循环嵌套用来从1开始计算，一直到n的阶乘数之和。其中内层循环的作用是将从1开始到n的每个数字的阶乘存放到变量k中。外层循环的作用是将每次内层循环结束时的k值累加到变量sum中，最后当外层循环结束时得到的sum值就是需要求的结果。

3.7 break语句

前面几节中已经讲述了C++的循环，包括while循环、do…while循环和for循环。这些循环根据自己的循环控制变量来控制循环的进程，但是在有些情况下，有时需要提前结束循环，这是循环控制变量做不到的。C++中提供了两种语句来完成此功能，这就是本节要讲述的break语句和下一节要讲述的continue语句。

前面在学习switch语句时用到了break语句。在switch语句中，break语句的作用是使程序跳出switch而去执行switch选择后面的语句。当break语句用于循环时，它的作用是使程序跳出当前循环，转而执行该循环后面的语句。break语句通常与if语句一起使用，意为：一旦满足什么条件则跳出。

break语句的一般形式如下：

```
for (表达式1; 表达式2; 表达式3) {
    …
    if (表达式) break;
    …
}
```

或者：

```
while (表达式1) {
    …
    if (表达式) break;
    …
}
```

实例3-21　求1到用户输入值的累加和。

```
#include <iostream.h>

void myFun();

void myFun(){
    short int i;
    short int n,sum;
    sum=0;

    cout<<" 请为变量 n 输入正整数值 n=";
    cin>>n;

    for (i=0;i<=n;i++) {
        sum = sum + i;
        if (sum < 0) {
            cout<<" 数值计算已超出 short int 型的范围! "<<"\n";
```

```
                    sum=-1;
                    break;
            }
        }
    cout<<"1 到 "<<n<<" 的和为 : "<<sum<<"\n";
}


main(){
    cout<<"*********************************"<<"\n";
    cout<<"*             求1到n的和            *"<<"\n";
    cout<<"*********************************"<<"\n";
    myFun();
    cout<<"*********************************"<<"\n";
}
```

运行程序，为变量n输入不同的数值，例如当n=100时，程序运行结果如下：

```
*********************************
*             求 1 到 n 的和           *
*********************************
请为变量 n 输入正整数值  n=100
1 到 100 的和为 : 5050
*********************************
```

当n=1000时，程序运行结果如下：

```
*********************************
请为变量 n 输入正整数值  n=1000
数值计算已超出 short int 型的范围!
1 到 1000 的和为 : -1
*********************************
```

通过上面两个结果可以看到，当为变量n赋值100时，因为所求和的取值范围没有超出short int型的取值范围，所以也不满足if条件，break语句不会运行。当为变量n赋值1000时，因为所求和的取值范围已经超出了short int型的取值范围，所以if语句将运行，在if语句中给出超范围提示后并运行break语句，强行退出for循环，转而运行for循环后面的语句。

需要指出的是，break语句强行退出的是本层循环体，在循环嵌套中，如果内层循环执行了break语句，则退出本层循环，而不会退出外层循环。

3.8 continue语句

break语句的作用是强行终止循环。本节将讲述continue语句，continue与break语句都是实现循环终止，所不同的是break语句是强行终止本层循环，使用break语句后，循环不再执行。而continue语句的作用是终止本次循环，使程序转到下一次循环。也就是说，使用continue语句后，循环并没有被终止，而是自动跳到下一次循环。

continue语句的一般形式如下：

```
for（表达式1；表达式2；表达式3）{
    …
    if（表达式）continue;
    …
}
```

或者：

```
while（表达式1）{
    …
    if（表达式）continue;
    …
}
```

下面通过一个例子来学习continue语句的用法。

实例3-22　求1到任意整数内所有能被3和5同时整除的数。

```cpp
#include <iostream.h>

void myFun();

void myFun(){
    int i,j;

    cout<<"请为变量j赋值 j=";
    cin>>j;

    for (i=1;i<=j;i++) {
        if (i%3==0 && i%5==0) {
            cout<<i<<"";
        }
        continue;
    }
    cout<<endl;

}
```

```
main(){
    cout<<"*******************************"<<"\n";
    cout<<"* 求 1 到任意整型数间能被 3 和 5 同时整除的数  *"<<"\n";
    cout<<"*******************************"<<"\n";
    myFun();
    cout<<"*******************************"<<"\n";
}
```

程序运行结果如下：

```
*******************************
*  求 1 到任意整型数间能被 3 和 5 同时整除的数  *
*******************************
请为变量 j 赋值  j=100
15  30  45  60  75  90
*******************************
```

在实例3-22中，在for循环内通过if…continue来判断哪些数能同时被3和5整除，并保持循环持续执行。if语句判断数值是否能被3和5同时整除，如果能被整除则输出这个数，否则就使用continue语句跳出本次循环，执行下一次循环。如此反复，直到完成整个循环。

3.9 goto语句

在C++中，还有一种语句，它的作用是使程序执行分支转移到被称为标号（label）的目的地，这种语句就是无条件转向语句。

实例 3-23 污水处理指标值验证程序。

```
#include <iostream.h>
#include <string.h>

void myFun();

void myFun(){
    float    exp1_value,exp2_value;
    float    result;
    label1:
        cout<<("请输入第一样本指数：");
        cin>>exp1_value;
        if ( exp1_value >9 || exp1_value <0)  {
```

```
            cout<<" 第一样本输入有误 (0-9), 请重新输入! "<<"\n";
            goto label1;
        }
    label2:
        cout<<" 请输入第二样本指数: ";
        cin>>exp2_value;
        if ( exp2_value >8 || exp2_value <0)  {
            cout<<" 第二样本输入有误 (0-8), 请重新输入! ";
            cout<<"\n";
            goto label2;
        }
        cout<<"\n";
        result = (exp1_value+exp2_value)/2;
        cout<<" 样本指数平均数  = "<<result<<"\n";
        if (result >=7)  {
            cout<<" 本次样品的检验结果为: 合格! ";
            cout<<"\n";
        }
        else  {
            cout<<" 本次样品的检验结果为: 不合格! ";
            cout<<"\n";
        }
}


main(){
    cout<<"**********************************\n";
    cout<<"*          污水指数验证模块         *\n";
    cout<<"**********************************\n";
    myFun();
    cout<<"**********************************\n";
}
```

程序运行结果如下:

```
**********************************
*          污水指数验证模块         *
**********************************
请输入第一样本指数: 10
```

```
    第一样本输入有误（0-9），请重新输入！
    请输入第一样本指数：9
    请输入第二样本指数：9
    第二样本输入有误（0-8），请重新输入！
    请输入第二样本指数：8


    样本指数平均数  = 8.5
    本次样品的检验结果为：合格！
    ****************************************
```

本例使用标号label1和label2来标识用户在输入的数值不在要求范围内的时候重新输入数值。goto语句的作用是在判断出用户输入的数值不在要求的范围内时，将程序跳转到label标识的位置。

使用goto语句时需要注意以下两点：

（1）标号不能单独出现，必须后跟一条语句或一条空语句。

（2）标号在其所在的函数内有效。

关于goto语句的使用在程序界有着广泛的争论。反对者认为：goto语句不符合结构化程序设计的思想，它让程序不容易理解，特别是往回调的goto语句，大大增加了程序结构的复杂性，给程序的排错带来很大的不便。支持者认为：goto语句使用非常灵活，在有些情况下（比如在多层循环嵌套中要从内层循环跳出），goto语句能够大大提高程序的效率。

关于这一点，笔者认为：应该在程序中谨慎使用goto语句，因为它在带来高效率的同时确实给程序的易读性和排错带来很大的不便。同时，在使用goto语句时也应该做相关的程序注释，尽量让程序特别是大型程序容易阅读，这样才容易发现和纠正程序的错误。

值得说明的是，在程序中，通常可以使用do…while语句来代替goto语句。关于这一点读者可以多看一些例题，并尝试将相关的goto语句用do…while语句代替，看程序运行的结果是不是跟使用goto语句一样。多做练习，你一定能够熟练掌握好的使用方法。

3.10 exit()和abort()函数

exit()函数在C++中的作用是退出当前程序，并且把程序的控制权返回给调用该程序的程序，括号里的返回值告诉调用程序，该程序的运行状态。不同的返回值表示不同的意义，例如：

- exit(0) 表示程序正常退出。
- exit(1) 表示程序出错退出。

实例3-24 求1到n的阶乘之和并判断用户输入是否为正数。

```cpp
#include <iostream.h>

#include <stulib.h>
```

```cpp
void myFun();

void myFun(){
    int i,j,n;
    float k,sum;
    char yN;

    label:cout<<" 请输入一个正整数值：";
    cin>>n;

    if (n<0)  {
        cout<<" 你输入的数值 "<<n<<" 是一个负数！"<<"\n";
        cout<<" 是否重新输入？ ";
        cin>>yN;
        if (yN=='y' || yN=='Y')
            goto label;
        else  {
            cout<<" 程序将退出！"<<"\n";
            exit(1);
        }
    }

    sum = 0;
    for (i=1;i<=n;i++)  {
        k = 1;
        for (j=1;j<=i;j++)
            k = k * j;
        sum = sum + k;
    }

    cout<<"1 到 "<<n<<" 的阶乘之和为：" <<sum<<"\n";
}

main(){
    cout<<"****************************"<<"\n";
    cout<<"*    求1!+2!+3!+...+n! 的值    *" <<"\n";
```

```
    cout<<"****************************"<<"\n";
    myFun();
    cout<<"****************************"<<"\n";
}
```

程序运行结果如下：

```
****************************
*    求 1!+2!+3!+...+n! 的值   *
****************************
请输入一个正整数值：-1
你输入的数值 -1 是一个负数！
是否重新输入？ y
请输入一个正整数值：-1
你输入的数值 -1 是一个负数！
是否重新输入？ Y
请输入一个正整数值：2
1 到 2 的阶乘之和为：3
****************************
```

实例3-24改写自实例3-20，在实例3-20的基础上增加了判断用户输入是否为正数的功能。如果用户输入正数，则执行结果与实例3-20的结果相同。否则，程序会询问用户是否重新输入，用户选择y或Y则程序通过goto语句跳转到输入数值的语句，否则程序执行exit(1)语句退出程序。

C++实现强行退出程序的还有一个abort()函数。exit()和abort()函数都是强行退出程序，所不同的是，exit()函数在退出程序时会完成一些必要的操作，比如释放程序中的变量、对象占用的内存、关闭打开的文件等。而abort()函数则不会做这些动作，只是直接退出程序。abort()函数多用于在程序遇到灾难性错误的时候强行退出程序。

实例3-25　求1到正整数n的和值并判断用户输入是不是正数。

```
#include <iostream.h>


void myFun();


void myFun(){
    short int i;
    short int n,sum;
    char yN;


    sum=0;
```

```
label:
    cout<<" 请为变量 n 输入正整数值 n=";
    cin>>n;
    if (n<0)  {
        cout<<" 你输入的 "<<n<<" 是一个负数 "<<"\n";
        cout<<" 是否要重新输入（Y/N）？ ";
        cin>>yN;
        if (yN=='y' || yN=='Y')
            goto label;
        else  {
            cout<<" 输入数值错误，程序异常退出！ "<<"\n";
            abort();
        }
    }

    for (i=0;i<=n;i++)  {
        sum = sum + i;
        if (sum < 0 )  {
            cout<<" 数值计算已超出 short int 型的范围！ "<<"\n";
            sum=-1;
            break;
        }
    }
    cout<<"1 到 "<<n<<" 的和为：" <<sum<<"\n";
}

main(){
    cout<<"*******************************"<<"\n";
    cout<<"*            求 1 到 n 的和            *"<<"\n";
    cout<<"*******************************"<<"\n";
    myFun();
    cout<<"*******************************"<<"\n";
}
```

程序运行结果如下：

```
*******************************

*            求 1 到 n 的和            *
```

```
* * * * * * * * * * * * * * * * * * * * * * * * * *
请为变量 n 输入正整数值  n=-1
你输入的 -1 是一个负数
是否要重新输入（Y/N）？ n
输入数值错误，程序异常退出！
```

实例3-25改写自实例3-21，程序的基本结构与实例3-24相似，请读者结合这两个例子自行分析，这里不再赘述。

3.11 对4种循环的比较

前面几节已经学习了C++的4种循环语句：while循环、do…while循环、for循环和goto语句。现在对这4种循环做一个简单的比较。

（1）在while循环和do…while循环的循环体中，必须有使循环趋向于结束的语句，而在for循环中的循环控制变量也必须使循环趋向结束，否则都将成为死循环。

（2）while循环和do…while循环在使用之前必须先初始化循环变量，而for循环则可以在表达式1中初始化循环变量。

（3）4种循环一般可以互相代替，但不建议使用goto语句。

（4）4种循环中的for循环最常用，功能也最为强大。

3.12 枚举类型

在实际生活中经常会遇到这样一类问题：一年包含12个月，一周有7天，一天有24个小时，人从性别上可以分为男人和女人等。这类问题反映在程序里可以用一个类型变量来表示，这种能够将变量的取值限定在一定范围之内的类型称为枚举类型。

定义枚举类型的一般形式如下：

```
enum 枚举名 {
    枚举值表
};
```

其中，枚举值表中应罗列所有可用值，这些值也称为枚举元素。例如：

```
enum human{man, woman};
```

该枚举名为human，枚举值共有两个，即男人和女人。凡被说明为human类型变量的取值都只能是男人或女人中的一个。

枚举变量可以采用不同的方式进行说明。

（1）先定义后说明。

```
enum human {man, woman};
enum human a, b;
```

（2）定义的同时加以说明。

```
enum human{man, woman} a, b;
```

（3）直接说明。

```
enum {man, woman} a, b;
```

枚举类型在使用中有如下规定：

（1）枚举值在枚举类型定义完成后已经确定，不能再对枚举值赋值。例如对上面定义的枚举变量human中的元素赋值。

```
man = 0;

woman = 1;
```

这种赋值都是错误的。

（2）枚举元素本身已由系统定义了一个表示序号的数值，从0开始顺序定义为0，1，2…例如在human中，man值为0，woman值为1。

（3）只能把枚举值赋予枚举变量，不能把元素的数值直接赋予枚举变量。

```
a=man;

b=woman;
```

上述赋值都是正确的，而以下这两种赋值则是错误的。

```
a=0;

b=1;
```

3.13　上机操作

比较两个数值的大小，将数值大的输出到屏幕上。

实例3-26　比较数值的大小。

```
#include <iostream.h>


void myFun();


void myFun(){
    short int a,b;
    char c;
    short int num;
    num=0;

    do  {
        num++;
        cout<<" 这是第 "<<num<<" 次比较 !"<<"\n";
```

```
            cout<<"a=";
            cin>>a;
            cout<<"b=";
            cin>>b;

            if (a>b)
                cout<<" 比较结果：a>b"<<"\n";
            else if (a==b)
                cout<<" 比较结果：a=b"<<"\n";
            else
                cout<<" 比较结果：a<b"<<"\n";

            cout<<" 你还要再比较其他数值吗 (Y/N)？ ";
            cin>>c;
    }while (c=='y' || c=='Y');
}


main(){
    cout<<"*********************************"<<"\n";
    cout<<"*            数值大小比较            *"<<"\n";
    cout<<"*********************************"<<"\n";
    myFun();
    cout<<"*********************************"<<"\n";
}
```

程序运行结果如下：

```
*********************************
*            数值大小比较            *
*********************************
这是第 1 次比较！
a= 1
b= 2
比较结果： a<b
你还要再比较其他数值吗 (Y/N)？ y
这是第 2 次比较！
a= 2
b= 1
```

比较结果：a>b

你还要再比较其他数值吗 (Y/N)？ y

这是第 3 次比较！

a= 1

b= 1

比较结果：a=b

你还要再比较其他数值吗 (Y/N)？ n

****

这个程序比较简单，用if…else循环来判断用户输入的两个值的大小，然后将比较结果输出到屏幕上。再用一个do…while循环来询问用户是否还要比较其他数值的大小。程序一开始就定义了一个short int型变量num，用来统计用户欲比较的次数。

用while循环求1～100的整数和。

　　实例3-27　用while循环求1～100的整数和。

```
#include <iostream.h>

void myFun();

void myFun(){
    int a,sum;

    a = 1;
    sum = 0;
    while (a<=100)  {
        sum = sum + a;
        a++;
    }
    cout<<"1+2+3+...+100="<<sum<<"\n";
}

main(){
    cout<<"****************************"<<"\n";
    cout<<"*    求 1+2+3+…+100 的值    *"<<"\n";
    cout<<"****************************"<<"\n";
    myFun();
    cout<<"****************************"<<"\n";
}
```

程序运行结果如下：

```
***************************
*    求 1+2+3+ … +100 的值    *
***************************
1+2+3+ … +100=5050
***************************
```

用do…while循环来求1～100的整数和。

实例3-28   用do…while循环求1～100的整数和。

```cpp
#include <iostream.h>

void myFun();

void myFun(){
    int a,sum;

    a = 1;
    sum = 0;
    do  {
        sum = sum + a;
        a++;
    }while (a<=100);
    cout<<"1+2+3+ … +100="<<sum<<"\n";
}


main(){
    cout<<"***************************"<<"\n";
    cout<<"*    求 1+2+3+ … +100 的值    *"<<"\n";
    cout<<"***************************"<<"\n";
    myFun();
    cout<<"***************************"<<"\n";
}
```

程序运行结果如下：

```
***************************
*    求 1+2+3+ … +100 的值    *
***************************
1+2+3+ … +100=5050
```

****

　　3.6.1节中用while循环求1+2+3+…+100的和与3.6.2节中用do…while循环求1+2+3+…+100的运行结果相同，其算法也是一样的，这里不再赘述。

　　下面是用for循环求1～100的整数和。

实例3-29　用for循环求1～100的整数和。

```
#include <iostream.h>

void myFun();

void myFun(){
  int a,sum;

  a = 1;
  sum = 0;

  for (;a<=100;)  {
    sum = sum + a;
    a++;
  }

  cout<<"1+2+3+…+100="<<sum<<"\n";
}

main(){
    cout<<"*************************"<<"\n";
    cout<<"*    求 1+2+3+…+100 的值    *"<<"\n";
    cout<<"*************************"<<"\n";
    myFun();
    cout<<"*************************"<<"\n";
}
```

程序运行结果如下：

****

*   求 1+2+3+…+100 的值   *

****

1+2+3+…+100=5050

****

这个例子与上面的两个例子从程序执行结果上看是一样的，程序中求和的算法也是一样的。在本例中，for循环的第一个循环变量表达式和第三个循环变量表达式省略了，而改在for循环的前面为循环变量a赋初值，在for循环的循环体中设置循环变量的自增量。

3.14 小结

本章讲述了C++的基本控制结构，包括顺序结构、选择结构、循环结构。这是C++组织程序的基础，其中顺序结构是最简单也是最基础的结构，选择结构又包括if语句和switch语句；循环结构包括while循环、do…while循环和for循环，其中for循环使用得最为频繁，在循环结构中还包括一个无条件转向的goto语句。

3.15 习题

一、填空题

1．程序语句是按照一定的顺序执行的，最简单的是按照_____顺序执行，这就是顺序结构。

2．C++的选择结构包括：_____语句、_____语句和_____语句。

3．C++的循环结构包括：_____语句、_____语句和_____语句。

4．循环嵌套的定义为：一个循环体内又包含_____，称为循环嵌套。如果内嵌的循环再嵌套_____，就称为多层嵌套。

5．C++中强行终止循环的语句包括：_____语句和_____语句。

二、编程题

编写一个程序，求1～100之间所有奇数的和。

三、程序阅读题

1．下面是一个求最小值的程序，请将该程序补充完整。

```cpp
#include <iostream.h>


void myFun();


void myFun(){
    short int a,b;
    cout<<"a=";
    _____;
    cout<<"b=";
    _____;
```

```cpp
    if (a>b)
       cout<<"min="<<b<<"\n";
    _____(a<b)
       cout<<"min="<<a<<"\n";
    _____ (a==b)
       cout<<"a=b"<<"\n";
}


main(){
    cout<<"**********************"<<"\n";
    cout<<"*        求最小值        *"<<"\n";
    cout<<"**********************"<<"\n";
    myFun();
    cout<<"**********************"<<"\n";
}
```

2．下面是一个倒序输出九九乘法表的程序，程序的运行结果是：

```
*************************************************************
*                 for 循环实现的九九乘法表                    *
*************************************************************
1*9=9 2*9=18 3*9=27 4*9=36 5*9=45 6*9=54 7*9=63 8*9=72 9*9=81

1*8=8 2*8=16 3*8=24 4*8=32 5*8=40 6*8=48 7*8=56 8*8=64

1*7=7 2*7=14 3*7=21 4*7=28 5*7=35 6*7=42 7*7=49

1*6=6 2*6=12 3*6=18 4*6=24 5*6=30 6*6=36

1*5=5 2*5=10 3*5=15 4*5=20 5*5=25

1*4=4 2*4=8 3*4=12 4*4=16

1*3=3 2*3=6 3*3=9

1*2=2 2*2=4

1*1=1

*************************************************************
```

请补充完整以下程序段：

```cpp
#include <iostream.h>


void myFun();


void myFun(){
```

```
    int sum = 0;
    cout<<"*************************************************************"<<"\n";
    cout<<"*                    for循环实现的九九乘法表                 *"<<"\n";
    cout<<"*************************************************************"<<"\n";
    for (short int a=___; a_____; a___)  {
        for (short int b=1;b<=a;b++)  {
            sum=a*b;
            cout<<b<<"*"<<a<<"="<<sum<<"";
            if (_____)
                cout<<"\n";
        }
    }
    cout<<"*************************************************************"<<"\n";
}


main(){
    myFun();
    return 0;
}
```

四、问答题

1. if语句与switch语句有什么区别?

2. while语句与do…while语句有什么区别?

3. 为什么说goto语句要谨慎使用?

4. exit()函数与abort()函数有什么不同?

第4章 C++函数

一个C++程序是由很多不同的功能模块组成的,每个功能模块都能够实现各自不同的功能,而这些功能又是通过一个个不同的函数来实现的。在一个C++程序中,可以用许多函数来完成不同的任务,但是在一个C++程序中有且只有一个主函数main()。主函数可以调用其他函数,函数之间也可以互相调用,而且这种调用并不限制次数。

4.1 主函数

main()函数是一个特殊的函数,它是C++程序中唯一可以直接执行的函数,其他函数都是通过直接或间接地被调用来执行的。

在一个简单的程序中,可以把程序功能具体实现的代码放在main()函数中,让main()函数来扮演功能函数的角色。但是在一个比较复杂的系统中,程序的总体任务被分割成不同的小任务,这些小任务都是由不同的功能函数来完成的;main()函数负责总体调度各个功能函数,而不做任何具体的功能实现。这一点在前几章的例子中已经体现出来了。

在函数调用中,不允许功能函数调用main()函数,因为这样做很容易出现死循环。

main()函数的一般形式如下:

```
main(){
    ...
    函数体
    ...
}
```

下面这个例子演示了在简单计算中main()函数的使用。

❀ 实例4-1 main()函数的使用。

```
#include <iostream.h>

main(){
```

```
short int a,b;  /* 定义变量 */
a=1;            /* 为变量赋初值 */
b=2;

cout<<"a+b="<<a+b<<"\n";  /* 求两个数值的和 */
return 0;
}
```

这个例子与前几章中看到的例子基本结构相同，而且比前面练习的例子更简单。只是需要注意以下几点：

（1）main()函数与其他函数一样，也有返回值。

（2）函数体以{开始，以}结束，花括号在使用中必须成对出现。

（3）C++中的注释以/*开始，以*/结束，开始与结束之间的部分为注释内容。

4.2 函数的定义

函数在形式上可以分为：无参函数和有参函数。

1．无参函数

这类函数没有形式参数，调用函数在调用它时不把参数传递给它，函数执行完成后可以把执行结果返回给主调函数，也可以不把执行结果返回给主调函数。无参函数定义的一般形式如下：

```
类型符  函数名() {
    声明部分
    语句
}
```

类型符说明了这个函数的类型，也就是这个函数返回值的类型。

2．有参函数

有参函数区别于无参函数，在函数声明时有一个形式参数列表。这类函数在被调用时需要按照形式参数列表给它传递相同顺序、相同类型、相同数目（可以少于参数数目，配对时按顺序进行）的参数，否则会引起调用错误。

函数执行完成后，可以把执行结果返回给主调函数，也可以不把执行结果返回给主调函数。有参函数定义的一般形式如下：

```
类型符  函数名 ( 形式参数列表 ) {
    声明部分
    语句
}
```

在前几章中没有接触到这类函数，下面通过一个例子来演示有参函数的声明、定义和使用方法。

实例4-2 利用无返回值有参函数计算两数相加。

```
#include <iostream.h>

void myFun1();
void myFun2(short int a,short int b);

void myFun1(){
    short int a, b;
    cout<<"a=";
    cin>>a;
    cout<<"b=";
    cin>>b;

    myFun2(a,b);
}

void myFun2(short int a,short int b){
    cout<<"a+b="<<a+b<<"\n";
}

main(){
    cout<<"****************************"<<"\n";
    cout<<"*    有参函数调用 - 无返回值    *"<<"\n";
    cout<<"****************************"<<"\n";
    myFun1();
    cout<<"****************************"<<"\n";
    return 0;
}
```

程序运行结果如下：

```
****************************
*    有参函数调用 - 无返回值    *
****************************
a=1
b=2
a+b=3
****************************
```

在实例4-2中，自定义函数myFun1是在前面几章中经常见到的函数形式。自定义函数myFun2与前面见到的函数不太一样，函数名后面的括号中不再是空的，而多了如下内容：

```
short int a,short int b
```

上面的类型定义放在函数名后面的括号中被称为形式参数列表，简称形参。这就意味着，当有函数要调用它时，必须把与形式参数列表中变量的顺序、变量类型、变量的数目相同的变量传递给它才可以调用，这些变量被称为实际参数，简称实参，否则将出现调用错误。

在自定义函数myFun1的函数体中有一条语句：

```
myFun2(a, b);
```

这条语句就是在调用自定义函数myFun2，并且把与myFun2的形式参数列表中的参数以相同顺序、相同类型、相同数目的参数传递给myFun2。

提示

在调用一个函数时，一定要注意它的形参列表（形参的顺序、类型、数量），在调用时也要根据形参列表来给出实参。否则，你可能调用的就不是这个函数了，因为C++中有重载函数，这些函数的函数名相同，但是函数的形参不同，实现的功能也不同，所以是不同的函数，这一点一定要注意。关于函数重载的概念和应用将在本章后面讲到。

在实例4-2中，自定义函数myFun1和myFun2前面都有一个void标识，这说明函数没有返回值。下面的例子是一个带有返回值的函数的示例。

实例4-3 利用有返回值的自定义函数编写的四则运算程序。

```cpp
#include <iostream.h>


void funInput();
char funSwitch(short int a,short int b);


void funInput(){
    short int a,b;

    cout<<"请输入变量 a 的值 :";
    cin>>a;
    cout<<"请输入变量 b 的值 :";
    cin>>b;
    cout<<"\n";

    switch (funSwitch(a,b))  {
        case'+': cout<<"a + b ="<<a+b<<"\n";break;
        case'-': cout<<"a - b ="<<a-b<<"\n";break;
        case'*': cout<<"a * b ="<<a*b<<"\n";break;
```

```
            case'/': cout<<"a / b ="<<a/b<<"\n";break;
            case'%': cout<<"a % b ="<<a%b<<"\n";break;
            default : cout<<"Error!"<<"\n";
        }
}


char funSwitch(short int a,short int b){
    char c;
    char d;
    cout<<"+"<<"\n";
    cout<<"-"<<"\n";
    cout<<"*"<<"\n";
    cout<<"/"<<"\n";
    cout<<"%"<<"\n";
    cout<<"请输入你要执行的计算：";
    cin>>c;

    switch (c)  {
        case'+': cout<<"你选择了加法操作："; d='+';break;
        case'-': cout<<"你选择了减法操作："; d='-';break;
        case'*': cout<<"你选择了乘法操作："; d='*';break;
        case'/':
          {
              if (b==0)
                  cout<<"除数不能为 0!"<<"\n";
              else  {
                  cout<<"你选择了除法操作：";
                  d='/';
              }
              break;
          }
        case'%':
          {
              if (b==0)
                  cout<<"除数不能为 0！ "<<"\n";
              else  {
```

```
                        cout<<" 你选择了求余操作 : ";

                        d='%';

                   }

               break;

           }

       default: cout<<" 你输入的字符不在选择范围内! "<<"\n";

   }

   return(d);

}


main(){

    cout<<"***************************"<<"\n";

    cout<<"*    有参函数调用 - 有返回值    *"<<"\n";

    cout<<"***************************"<<"\n";

     funInput();

    cout<<"***************************"<<"\n";

     return 0;

}
```

程序运行结果如下:

```
***************************

*    有参函数调用 - 有返回值    *

***************************

请输入变量 a 的值 : 1

请输入变量 b 的值 : 2


+

-

*

/

%

请输入你要执行的计算 : /

你选择了除法操作 : a  /  b  =  0

***************************
```

　　这个例子看上去有点复杂，但其实这个例子在前面的实例3-8中已经看到过了，只不过那时使用了一个无返回值的自定义函数来完成主要的功能。在本例中使用了两个自定义函数funInput()和funSwitch()，其中funSwitch()函数是一个带参数和返回值的自定义函数，它的函数原型如下：

```
char funSwitch(short int a,short int b);
```

这个自定义函数的形参列表中有两个short int类型的变量，这就要求调用它的程序也必须传递给它两个short int类型的变量作为实参。同时，这个函数是一个char型的函数，这意味着该函数可以向调用它的程序返回一个char类型的值。向主调程序返回值的语句一般形式为return(返回值)。这个返回值的类型必须和函数的类型一致。

4.3 局部变量

C++中的变量都有其作用域，根据变量定义的位置不同，其作用域也不同。据此，可以将C++中的变量分为局部变量和全局变量。

在一个函数的内部定义的变量，只在本函数内部有效，也就是其作用域在本函数内，这种变量称为局部变量。例如：

```
int myFun1(char b){
  …
  short int a;
  …
}
```

还有：

```
int myFun2(){
  …
  char a;
  …
}
```

上面定义了两个自定义函数myFun1和myFun2。在函数myFun1中定义的形参b和变量a都只在本函数内有效，而在函数myFun2中无效。在函数myFun2中定义的变量a也只在本函数内有效。在这两个函数中，虽然都有变量a，但是它们没有任何关系，它们是完全独立的。实例4-4演示了局部变量的作用域。

实例4-4　局部变量的作用域。

```
#include <iostream.h>

void myFun1();
short int myFun2(short int b);

void myFun1(){
    short int a;
```

```
        cout<<"myFun1 的变量：a=";

        cin>>a;


        cout<<"myFun1 的变量 a+myFun2 的变量 b="<<myFun2(a)<<"\n";
    }

short int myFun2(short int b){
        short int a;
        short int c;

        cout<<"myFun2 的变量：a=";
        cin>>a;
        c=a+b;
        return(c);
    }


main(){
        cout<<"*******************************"<<"\n";
        cout<<"*          局部变量的作用域          *"<<"\n";
        cout<<"*******************************"<<"\n";
        myFun1();
        cout<<"*******************************"<<"\n";
        return(0);
    }
```

程序运行结果如下：

```
*******************************
*          局部变量的作用域          *
*******************************
myFun1 的变量：a=1
myFun2 的变量：a=2
myFun1 的变量 a+myFun2 的变量 b=3
*******************************
```

在本例中定义了两个自定义函数myFun1和myFun2，其中函数myFun2带有参数和返回值。在自定义函数myFun1中定义了一个short int类型的变量a，并要求用户为这个变量赋值。

在自定义函数myFun2中也定义了一个short int类型的变量a，也要求用户为这个变量赋值。

当函数myFun1调用函数myFun2的时候，按照函数myFun2的要求向myFun2传递一个short int类型的变量，实际上就是把myFun1中的变量a的值赋给myFun2中的形参b，然后这两个值在函数myFun2中相加，再通过return语句把相加结果传给myFun2的主调函数myFun1。从程序的运行结果可以清楚地看到局部变量的作用域。

4.4 全局变量

除了第4.3节讲到的局部变量外，C++中还有一种变量，这种变量的定义不在任何函数内部，所以它不属于任何一个函数私有，而属于所有函数公用，这种变量称为全局变量或外部变量。全局变量的作用范围是从变量的定义位置到整个源程序文件的结束。例如：

```
int a,b;
myFun1(){
    …
}


myFun2(){
    …
}
```

这里定义的变量a和b是全局变量，它们的作用范围是从int a,b;语句到源程序的结束处。变量a和b在函数myFun1和myFun2中都同样有效。

实例4-5 全局变量的使用。

```
#include <iostream.h>

void myFun1();
void myFun2();

short int num;
void myFun1(){
    short int a;

    cout<<"myFun1 的变量 : a=";
    cin>>a;

    cout<<"myFun1 的变量 a+ 全局变量 num="<<a+num<<"\n";
}
```

```
void myFun2(){
    short int a;

    cout<<"myFun2 的变量：a=";
    cin>>a;
    cout<<"myFun2 的变量 a+ 全局变量 num="<<a+num<<"\n";
}


main(){
    cout<<"*******************************"<<"\n";
    cout<<"*            全局变量的作用域          *"<<"\n";
    cout<<"*******************************"<<"\n";
    num=100;
    myFun1();
    myFun2();
    cout<<"*******************************"<<"\n";
    return(0);
}
```

程序运行结果如下：

```
*******************************
*           全局变量的作用域         *
*******************************
myFun1 的变量：a=1
myFun1 的变量 a+ 全局变量 num=101
myFun2 的变量：a=2
myFun2 的变量 a+ 全局变量 num=102
*******************************
```

在本例中定义了一个short int型的全局变量num，然后在主函数中为这个变量赋初值。在自定义函数myFun1和myFun2中，分别调用这个全局变量与各自的局部变量相加，最后输出结果。通过该例子可以看到，全局变量在整个程序文件中都有效，所有函数（包括主函数）都可以自由地使用它。

全局变量的作用域是从变量定义的位置到程序文件的结束处，而在实例4-5中在程序的开始处定义了全局变量，那么它下面的函数自然都可以使用这个全局变量。如果把全局变量的定义移到函数后面，情况会怎样呢？下面是这个变量定义的位置：

```
#include <iostream.h>
```

```
void myFun1();

void myFun2();


void myFun1(){

   …

}


short int num;


void myFun2(){

   …

}


main(){

   …

}
```

如果在这个位置定义全局变量，则在编译时编译器会报一个错误：全局变量num在函数中没有定义。这符合全局变量的作用域的范围。

要解决这个问题，需要在函数中声明这个全局变量，声明全局变量时我们使用了说明符extern，其一般形式如下：

extern 变量类型 变量名

通常情况下，在函数中使用全局变量需要使用extern声明，如果函数的定义是在源程序的开始位置（即在所有使用全局变量的函数之前），那么这步声明可以省略。加上全局变量的声明后，再来看看程序的形式。

实例4-6　全局变量在函数中的声明。

```
#include <iostream.h>


void myFun1();

void myFun2();


void myFun1(){

    extern short int num;


    short int a;


    cout<<"myFun1 的变量：a=";
```

```
        cin>>a;

        cout<<"myFun1 的变量 a+ 全局变量 num="<<a+num<<"\n";
    }

    void myFun2(){
        extern short int num;

        short int a;

        cout<<"myFun2 的变量 : a=";
        cin>>a;
        cout<<"myFun2 的变量 a+ 全局变量 num="<<a+num<<"\n";
    }

    main(){
        extern short int num;

        cout<<"*******************************"<<"\n";
        cout<<"*      全局变量在函数中的声明      *"<<"\n";
        cout<<"*******************************"<<"\n";
        num=100;
        myFun1();
        myFun2();
        cout<<"*******************************"<<"\n";
        return(0);
    }
```

程序运行结果如下：

```
*******************************
*      全局变量在函数中的声明      *
*******************************
myFun1 的变量 : a=1
myFun1 的变量 a+ 全局变量 num=101
myFun2 的变量 : a=2
myFun2 的变量 a+ 全局变量 num=102
```

****

在本例中把全局变量的声明放在程序文件的最后，使用这种方法时必须在每个函数内部使用extern来声明这个全局变量，然后才可以在函数中使用该全局变量，否则会引起编译错误。

关于全局变量和局部变量需要做如下说明：

(1) 局部变量的作用域仅限于该变量所在的函数内部，全局变量的作用域则是整个程序文件。

(2) 局部变量在其所属函数内部定义和使用，而全局变量必须在函数外部定义，在使用该全局变量的函数中声明。全局变量在定义时就已经分配了内存单元，同时可以赋予其初值，但是在函数内部说明时不能再为其赋初值。

(3) 因为全局变量可以在所有函数中共用，所以全局变量的存在加强了各函数之间的联系。但是，这样却带来了另外一个问题，即模块化程序设计强调各功能函数的独立性，而全局变量则使这种独立性降低。从一定程度上讲，全局变量不符合模块化程序设计的思想。因此，在不是必须使用全局变量的时候，尽量不要使用全局变量。

(4) 在同一个源文件中，全局变量和局部变量可以同名，在具体函数中，如果有同名的局部变量和全局变量，那么全局变量则不起作用。

下面这个例子说明了在同一个函数中有重名的全局变量和局部变量的情况。

实例4-7 在一个函数中有同名的全局变量和局部变量。

```
#include <iostream.h>

void myFun1();
void myFun2();

short int a;

void myFun1(){
    short int a;

    a = 10;
    cout<<" 局部变量有效：a = "<<a<<"\n";
}

void myFun2(){
    cout<<" 全局变量有效：a = "<<a<<"\n";
}

main(){
    a = 20;
```

```
        cout<<"******************************"<<"\n";
        cout<<"*      全局变量与局部变量同名      *"<<"\n";
        cout<<"******************************"<<"\n";
        myFun1();
        myFun2();
        cout<<"******************************"<<"\n";
        return 0;
}
```

程序运行结果如下：

```
******************************
*      全局变量与局部变量同名      *
******************************
局部变量有效：a = 10
全局变量有效：a = 20
******************************
```

在本例中定义了一个全局变量a，a变量与自定义函数myFun1中的一个局部变量重名，而自定义函数myFun2中没有局部变量。从程序运行结果来看，在自定义函数myFun1中，起作用的是局部变量a，而在自定义函数myFun2中起作用的则是全局变量a。

> **提示**
>
> 全局变量在定义完成后，在整个程序的执行期间，其所占据的空间都不会被释放，这种变量有悖于面向对象的思想，应尽量少用全局变量。

4.5 变量的存储类型

通过前面两节的内容可以知道不同的变量有着不同的作用域：局部变量的作用域仅限于声明它的函数内部，而全局变量的作用域则是整个程序文件。事实上，不同的变量在内存中的存储类型是不同的。所谓存储类型就是指变量占用内存空间的方式，也称为存储方式。

变量的存储方式可以分为静态存储和动态存储两种。静态存储变量是在定义变量时系统就为它分配了内存单元，并且在程序运行期间一直不变，直至整个程序运行结束才释放这些内存空间。而动态存储变量则是在程序运行中使用到这些变量时系统才为其分配内存单元，使用完毕后这些内存单元马上被释放。

C++中定义的变量的存储类型有4种：自动变量（auto）、外部变量（extern）、寄存器变量（register）和静态变量（static）。

在前面已经学习过变量的定义方法：数据类型说明符 变量名，变量名，…现在又学习了变量的存储类型，因此可以给出变量定义的完整形式如下：

存储类型说明符　数据类型说明符　变量名，变量名，…；

例如：

```
auto short int a,b; /* 说明变量a、b是short int 类型的自动变量 */
extern char a,b; /*说明变量a、b是char 类型的外部变量 */
static float a,b; /* 说明变量a、b是float 类型的静态变量 */
```

下面对这几种变量的存储类型逐一说明。

1. 自动变量

这种变量使用得最为普遍。凡是在变量定义时未加存储类型说明符的变量均为自动变量。自动变量就是在函数内部定义的变量。它只允许在定义它的函数内部使用，而在函数外的其他任何地方都不能使用。

在前面几章所使用的例子中定义的变量基本上都是自动变量（除本章中定义过一些extern类型的变量之外）。例如：

```
{
…
short int a;
char b;
float c;
…
}
```

这种定义方式与下面的定义方式是等价的：

```
{
…
auto short int a;
auto char b;
auto float c;
…
}
```

自动变量是局部变量，即它的作用域是在定义它的函数内部有效。当然，这说明自动变量也没有链接性，因为它也不允许其他文件访问它。

由于自动变量在定义它的函数外的任何地方都不可见，所以允许在这个函数外的其他地方或者其他函数内定义同名的变量，它们之间不会发生冲突。因为它们都有自己的作用域，而且没有链接性（即不允许其他文件访问它）。

对于自动类型的变量要注意的是：自动变量属于动态存储方式，只有在使用时它才会由系统分配内存单元，使用完毕后会立即释放其所占用的内存单元。因此，在一个函数内定义的自动变量，只有当这个函数被调用时才会被分配内存单元，在函数调用结束后立即释放其所占用的内存单元，这就是自动变量的持续性。计算机在执行这个函数时，创建并为它分配内存，当函数执行完毕返回后，自动变量就会被销毁，这个过程是通过堆栈机制来实现的。为自动变量分配内存就是进栈，而函数返回时就是出栈。

在复合语句中也是一样的，例如下面这个函数：

```
void myFun{
    short int a;
    …
    {
        short int b;
        …
    }
}
```

在上面这个函数中，变量a在定义时省略了auto修饰符，它是一个自动变量，只在本函数内有效。在这个函数中，包含一个复合语句，在这个复合语句中又定义了一个auto变量b，该变量也只在这个复合语句中有效。

另外，如果在一个函数内部有复合语句，那么该函数中的自动变量可以与复合语句中的自动变量同名，这时在复合语句内定义的变量是有效的。

下面这个例子说明了自动变量的作用域。

实例4-8　自动变量的作用域。

```
#include <iostream.h>

void myFun();

void myFun(){
    auto short int a;
    auto short int b,c;
    b=1;
    c=2;

    cout<<"a=";
    cin>>a;

    if (a==0)  {
        auto short int a;
        auto short int b,c;

        b=3;
        c=4;
```

```
                    cout<<" 复合语句内 : a=b+c="<<b+c<<"\n";
        }

        cout<<" 复合语句外 : a=b+c="<<b+c<<"\n";
}


main(){
    cout<<"****************************"<<"\n";
    cout<<"*      自动变量的作用域       *"<<"\n";
    cout<<"****************************"<<"\n";
    myFun();
    cout<<"****************************"<<"\n";
}
```

程序运行结果如下：

```
****************************
*       自动变量的作用域       *
****************************
a=0
复合语句内 : a=b+c=7
复合语句外 : a=b+c=3
****************************
```

本例中首先在自定义函数myFun中定义了3个auto类型的变量a、b和c，其中变量b和c在后面的复合语句中还有同名定义。根据程序运行的结果可以看到自动变量的作用域。

请读者自行分析以下代码：

```c
#include <stdio.h>

int main(){
    int print();
    int var,i;
    for (i=0; i<=10; i++)
      var=print();
    printf("%d\n",var);
    return 0;
}
```

```
int print(){
    auto int i=0;  /* 自动变量 */
    i+=1;
    printf("%d\n",i);
    return i;
}
```

2．外部变量

外部变量其实就是前面说的全局变量（见第4.4节），这里不再赘述。

3．静态变量

静态变量的类型说明符是static，静态变量当然属于静态存储方式，但是属于静态存储方式的变量不一定就是静态变量，例如外部变量虽属于静态存储方式，但不一定是静态变量，必须由static加以定义后才能成为静态外部变量，或称静态全局变量。对于自动变量，前面已经介绍过它属于动态存储方式。但是也可以用static定义它为静态自动变量，或称静态局部变量，从而成为静态存储方式。

由此看来，一个变量可以由static进行再说明，并改变其原有的存储方式。

A．静态局部变量

在局部变量的说明前再加上static说明符就构成静态局部变量。例如：

```
static int a, b;
static float array[5]={1, 2, 3, 4, 5};
```

静态局部变量属于静态存储方式，它具有以下特点：

（1）静态局部变量在函数内定义，但不像自动变量那样，当调用时就存在，退出函数时就消失。静态局部变量始终存在着，也就是说它的生存期为整个源程序。

静态变量与自动变量的本质区别是，静态变量并不像自动变量那样运用堆栈机制来使用内存，而是为静态变量分配固定的内存。在程序运行的整个过程中，它都会被保持，而不会被销毁，就是说静态变量的持续性是程序运行的整个周期。

这有利于我们共享一些数据，如果静态变量在函数内部定义，则它的作用域就是在这个函数内部，即仅在这个函数内部使用它才有效。但是它不同于自动变量，自动变量离开函数后就会被销毁，而静态变量不会被销毁。它在函数的整个运行周期内都会存在。

在函数外定义的变量为全局变量，工程内的所有文件都可以访问它，但是它在整个工程内只能定义一次，不能有重复定义，不然就会发生错误，而其他文件要想使用这个变量，则必须用extern来声明这个变量，这个声明叫做引用声明。这一点很重要，如果没有用extern来声明已经在其他文件中定义的全局变量就使用它，则会发生错误。

如果只想在定义它的文件中使用它，而不允许在其他文件中使用它，那么就用关键字static在函数外部声明变量。这样这个变量在其他文件中将不可见，即它的链接性只是内部链接。有一点是需要注意的，如果在函数外部声明一个变量：

```
const int a;
```

那么变量a的链接性为内部链接，只能在定义它的文件内使用。还有，如果在定义静态变量时并没有赋予变量初始值，那么静态变量将被自动初始化为0。

（2）静态局部变量的生存期虽然为整个源程序，但是其作用域仍与自动变量相同，即只能在定义该变量的函数内部使用该变量。退出该函数后，尽管该变量还继续存在，但不能再使用它。

（3）允许对构造类静态局部量赋初值。在介绍数组初始化时已作过说明，若未赋予初值，则由系统自动赋予初值0。

（4）对于基本类型的静态局部变量，若在说明时未赋予初值，则系统自动赋予初值0。若对自动变量不赋初值，则其值是不定的。

根据静态局部变量的特点，可以看出它是一种生存期为整个源程序的变量。虽然离开定义它的函数后不能再使用，但再次调用定义它的函数时，它又可继续使用，而且保存了前次被调用后留下的值。因此，当多次调用一个函数且要求在调用之间保留某些变量的值时，可考虑采用静态局部变量。虽然使用全局变量也可以达到这种目的，但全局变量有时会造成意外的副作用，所以仍以采用局部静态变量为宜。

```
main() {
    int i;
    void f(); /* 函数说明 */
    for (i=1; i<=5; i++)
        f(); /* 函数调用 */
}
void f() {/* 函数定义 */
    auto int j=0;
    ++j;
    printf("%d\n",j);
}
```

上述程序中定义了函数f，其中的变量j说明为自动变量并赋予初始值0。当main中多次调用f时，j均赋初值为0，故每次输出值均为1。现在把j改为静态局部变量，程序如下：

```
main(){
    int i;
    void f();
    for (i=1; i<=5; i++)
        f();
}
void f() {
    static int j=0;
    ++j;
    printf("%d\n",j);
}
}
```

由于j为静态变量，能在每次调用后依然保留其值并在下一次调用时继续使用，所以输出值成为累加的结果。读者可自行分析其执行过程。

B．静态全局变量

在全局变量（就是外部变量）的说明之前再冠以static就构成了静态全局变量。全局变量本身就是静态存储方式，静态全局变量当然也是静态存储方式。两者在存储方式上并无不同，区别在于非静态全局变量的作用域是整个源程序。当一个源程序由多个源文件组成时，非静态全局变量在各个源文件中都有效；而静态全局变量则限制了其作用域，即只在定义该变量的源文件内有效，在同一源程序的其他源文件中不能再使用它。

由于静态全局变量的作用域局限于一个源文件内，只能为该源文件内的函数公用，因此可以避免在其他源文件中引起错误。从以上分析可以看出，把局部变量改变为静态变量后是改变了它的存储方式，即改变了它的生存期。而把全局变量改变为静态变量后则改变了它的作用域，限制了它的使用范围。因此static修饰符在不同的地方所起的作用不同，应予以注意。

```c
#include <stdio.h>


int main(){
    int print();
    int var,i;
    for (i=0; i<=10; i++)
        var=print();
    printf("%d",var);
    return 0;
}


int print(){
    static int i;   /*静态变量 */
    i+=1;
    printf("%d\n",i);
    return i;
}
```

以后在第5章中还将具体举例说明。

4．寄存器变量

上述各类变量都存放在存储器内，因此当对一个变量频繁读写时，必须反复访问内存储器，从而花费大量的存取时间。为此，C++语言提供了另一种变量，即寄存器变量。这种变量存放在CPU的寄存器中，使用时不需要访问内存，而直接从寄存器中读写，这样可提高效率。

寄存器变量的修饰符为register。对于循环次数较多的循环控制变量及循环体内反复使用的变量均可定义为寄存器变量。

实例4-9 求1～200之间的整数之和。

```
main(){
    register i,s=0;
    for (i=1; i<=200; i++)
        s=s+i;
    printf("s=%d\n",s);
}
```

本程序循环次数为200次，i和s都将频繁使用，因此可定义为寄存器变量。对寄存器变量还要说明以下几点：

（1）只有局部自动变量和形参才可以定义为寄存器变量，全局变量不能说明成寄存器变量，即在函数外声明的变量不能使用register修饰符。因为寄存器变量属于动态存储方式，凡需要采用静态存储方式的变量都不能定义为寄存器变量。

（2）在Turbo C，MS C等版本所使用的C语言中，实际上是把寄存器变量当成自动变量处理，因此速度并不见得提高。而在程序中允许使用寄存器变量只是为了与标准C保持一致。

（3）即使能真正使用寄存器变量的机器，由于CPU中寄存器的个数有限，因此使用寄存器变量的个数也是有限的。

在C语言中可以使用寄存器变量来优化程序的性能，最常见的就是在一个函数体当中，将一个常用的变量声明为寄存器变量，例如：

```
register int ra;
```

设置寄存器变量的目的是提高对有关变量的存取速度，存取寄存器的速度要比存取内存单元快得多。例如，一个循环语句的控制变量可以声明为寄存器变量，一般变量不能说明为寄存器变量。

寄存器变量采用高效寻址方式，而且寄存器的速度远远高于CPU二级缓存的速度，更高于内存的速度，从而优化程序的运行速度。但寄存器的数量很少，一般不超过100个，还不到1KB，相对于动辄数GB的内存来说，少得可怜，不过大部分的指令都是需要寄存器的参与的，在C中不能对寄存器变量进行取址，而在C++中是可以的，因为这时它已经成为一个普通的内存对象。

通常寄存器的地址不用存取，因此不管一个寄存器变量实际是否已分配在寄存器中，都不在程序中使用它的地址，所以不让指针指向寄存器变量。例如：

```
register char C;
char *cp=&C;  // 错误
```

再看下面的例子：

```
#include <iostream>

using namespace std;

int main(void){
```

```
    register int iVar;
    int *piVar;

    iVar = 10;
    piVar = &iVar;
    cout <<"*piVar =" << *piVar << endl;

    return 0;
}
```

只要使用了register修饰符，那么对该变量进行取地址的操作就是被禁止的，不管事实上该变量是否被放入寄存器中。被register修饰的变量称为寄存器变量，它只能用于整型和字符型变量。修饰符register定义的变量在Turbo C里存储在CPU寄存器中，而不是像普通变量那样存储在内存中，这样可以提高运算速度。但是Turbo C中只允许同时定义两个寄存器变量，一旦超过两个，那么编译程序时就会自动将超过限制数目的寄存器变量当作非寄存器变量来处理。因此，寄存器变量常用在同一变量名频繁出现的地方。

4.6 函数的调用

在前面几章的例子中已经使用过函数调用，几乎在所有实例中都要定义自定义函数，然后从主函数中调用这些函数。

函数调用的一般形式如下：

函数名（实参列表）

在前面学习有参函数时已经谈到了有参函数的调用。对于函数的调用需要注意以下两点：

（1）函数调用分为无参函数调用和有参函数调用。在无参函数调用时没有实际参数，而在有参函数调用时是有实际参数的，因此实参的顺序、数据类型和数量与被调函数中形参的顺序、数据类型和数量（可以少于）必须一致。

（2）实参列表中的参数可以是常数、变量或表达式。

```
myFun(a,b);
myFun1(1,b);
myFun(a,a+b)
```

下面是一个函数调用的例子。

实例4-10　求学生的总成绩和平均成绩。

```
#include <iostream.h>
#include <math.h>

void myFun();
```

```
float calcAvgFun(float a,float b,float c);
float calcSumFun(float a,float b,float c);

void myFun(){
    float chinese,math,english;
    float sumresult,avgresult;

    cout<<" 请录入语文成绩:";
    cin>>chinese;
    cout<<" 请录入数学成绩:";
    cin>>math;
    cout<<" 请录入英语成绩:";
    cin>>english;

    sumresult=calcSumFun(chinese,math,english);
    cout<<" 总成绩为:"<<sumresult<<"\n";
    avgresult=calcAvgFun(chinese,math,english);
    cout<<" 平均成绩:"<<avgresult<<"\n";
}

/* 求平均成绩的函数 */
float calcAvgFun(float a,float b,float c){
    float avgabc;

    avgabc = (a+b+c)/3;
    return(avgabc);
}

/* 求总成绩的函数 */
float calcSumFun(float a,float b,float c){
    float sumabc;

    sumabc = a+b+c;
    return(sumabc);
}
```

```
main(){
    cout<<"*********************************"<<"\n";
    cout<<"*     求学生的总成绩和平均成绩     *"<<"\n";
    cout<<"*********************************"<<"\n";
    myFun();
    cout<<"*********************************"<<"\n";
}
```

程序运行结果为：

```
*********************************
*      求学生的总成绩和平均成绩      *
*********************************
请录入语文成绩:95.5
请录入数学成绩:96.5
请录入英语成绩:97.5
总成绩为:289.5
平均成绩:96.5
*********************************
```

本例中使用了两个有参函数calcSumFun和calcAvgFun来分别求总成绩和平均成绩，用一个无参函数myFun来实现成绩输入功能，并调用calcSumFun和calcAvgFun函数。calcSumFun和calcAvgFun是两个带有返回值的函数，函数myFun在调用这两个函数完成后将通过局部变量sumresult和avgresult分别接受两个函数的返回值，即总成绩和平均成绩。

另外，也可以不使用变量来接收函数的返回值，而是直接使用函数调用的表达式来得到函数的返回值。

```
void myFun(){
    cout<<"请录入语文成绩:";
    cin>>chinese;
    cout<<"请录入数学成绩:";
    cin>>math;
    cout<<"请录入英语成绩:";
    cin>>english;

    cout<<"总成绩为:"<<calcSumFun(chinese,math,english)<<"\n";
    cout<<"平均成绩:"<< calcAvgFun(chinese,math,english)<<"\n";
}
```

读者可以把修改后的函数替换实例4-9中的函数myFun，然后查看运行结果与实例4-9是否一致。

4.7 函数的递归调用

函数的递归调用指一个函数在它的函数体内，直接或间接地调用它自身。由于函数的递归调用是在调用自己，因此在递归调用中必须有方法避免函数无休止地运行，即出现死循环的情况。

通常情况下使用条件判断可以解决这个问题，即在满足一定条件下，可以进行递归调用，当条件不能满足时即终止递归调用。

下面这个例子演示了函数的递归调用。

实例4-11　用递归法求1*2*3*…*n的值。

```
#include <iostream.h>

short int myFun(short int n);

short int myFun(short int n){
    if(n==0)
      return 1;
    else
    return n*myFun(n-1);
}

main(){
    cout<<"**************************"<<"\n";
    cout<<"* 用递归函数求1到n的乘积 *"<<"\n";
    cout<<"**************************"<<"\n";
    short int a;

    cout<<"a=";
    cin>>a;
    cout<<"1*…*"<<a<<"="<<myFun(a)<<"\n";
    cout<<"**************************"<<"\n";
}
```

程序运行结果如下：

```
**************************
* 用递归函数求1到n的乘积 *
**************************
a=6
```

```
1*…*6=720

***************************
```

在这个例子中定义了一个返回值为short int类型的函数myFun，通过对这个函数的递归调用来求解1*2*3*…*n的值。为了防止函数无休止地递归调用下去，设立了一个判断条件if(n==0)，即从n开始乘，一直乘到1为止。

下面再来演示一个C++经典的递归调用的例子——汉诺塔问题。

汉诺塔问题来源于印度一个古老的传说：在印度的一个神庙中有3根金刚石的棒，第一根上面套着64个圆形金盘，最大的在下面，其余的按照大小逐个往上排，最小的放在最上面，庙里的僧人把它们一个个地从这根棒转移到另一根棒上，按照规定可以利用中间的一根棒作为桥梁，但每次只能搬一个，而且要始终保持大的在小的下面。

汉诺塔问题可以转化为下面的游戏：

（1）有三根杆子a、b和c。其中a杆上有64个圆盘，它们按照大小顺序从上到下排列，最大的被放在最下面，然后逐个往上排，最小的被放在最上面。

（2）把所有圆盘由a杆移动到c杆上，每次只能移动一个。

（3）在移动过程中要保持大的在下小的在上的原则。

在这个问题中，由于条件限制每次只能移动一个圆盘，而且不允许大圆盘放在小圆盘之上。

经过计算，要移动这64个圆盘需要的次数是18 446 744 073 709 551 615次！这是一个天文数字，以目前的计算机技术很难解决64层汉诺塔问题，所以只能通过解决较小层数的汉诺塔问题来说明解决方案，以下是通过函数的递归调用解决汉诺塔问题的程序。

实例4-12　求解汉诺塔。

```cpp
#include <iostream.h>


void hanota(short int diskNum,short int FromRod,short int ByRod,short int ToRod);

void hanota(short int diskNum,short int FromRod,short int ByRod,short int ToRod){
    if (diskNum==1)
      cout<<FromRod<<"->"<<ToRod<<"\n";
    else {
      hanota(diskNum-1,FromRod,ToRod,ByRod);
      cout<<FromRod<<"->"<<ToRod<<"\n";
      hanota(diskNum-1,ByRod,FromRod,ToRod);
    }
}


main(){
    short int diskNum;
```

```
        cout<<"**************************"<<"\n";
        cout<<"*   递归法解决汉诺塔问题  *"<<"\n";
        cout<<"**************************"<<"\n";
        cout<<" 请输入汉诺塔的层数 : ";
        cin>>diskNum;
        cout<<" 步骤如下 : "<<"\n";
        hanota(diskNum,1,2,3);
        cout<<"**************************"<<"\n";
}
```

程序运行结果如下：

```
**************************
*   递归法解决汉诺塔问题  *
**************************
请输入汉诺塔的层数 : 3
步骤如下 :
1->3
1->2
3->2
1->3
2->1
2->3
1->3
**************************
```

程序运行结果是，每次移动圆盘的具体步骤为：先从第一根杆子上移动1个盘子到第三根杆子，第二步从第一根杆子上移动1个盘子到第二根杆子上。根据运行结果依次类推，直到最后把所有的盘子都移动到第三根杆子上为止。

需要指出的是，当汉诺塔的层数足够多时，利用普通计算机几乎不可能完成求解，因此读者在运行这个程序时尽量不要输入较大的值。

实例4-13　用递归法求n!。

```
#include <iostream.h>

void inputNum();
long exponential(short int n);

void inputNum(){
    short int m;
```

```
        cout<<" 请输入要求阶乘的数 :";
        cin>>m;
        cout<<m<<"! ="<<exponential(m)<<"\n";
}


long exponential(short int n){
    long result;

    if(n>1)
      result=n*exponential(n-1);
    else
      result=1;


    return(result);
}


main(){
    cout<<"*****************************"<<"\n";
    cout<<"*        用递归法求阶乘       *"<<"\n";
    cout<<"*****************************"<<"\n";
    inputNum();
    cout<<"*****************************"<<"\n";
}
```

程序运行结果如下：

```
*****************************
*        用递归法求阶乘       *
*****************************
请输入要求阶乘的数 :5
5! =120
*****************************
```

在本例中定义了两个自定义函数inputNum()和long exponential(short int n)，函数inputNum用来实现接收用户输入的要求阶乘的数，函数exponential利用递归调用求阶乘运算。

程序从main()函数开始执行，由main()调用inputNum()函数，后者再调用函数exponential。函数之间这种调用关系就是下一节将要讲的函数的嵌套调用。

4.8 函数的嵌套调用

C++的源程序是由多个功能函数组成的，这些功能函数彼此之间是平等的关系，它们都能被main()函数调用，同时它们之间也能够互相调用。在一个函数调用另外一个函数时，被调函数还可以再调用其他函数，称为函数的嵌套调用。在4.7节中学习的函数的递归调用，那是函数嵌套调用的一个特例，即一个函数在被调用的同时又在调用自己。

前面已经讲过，C++程序中有且只有一个main()函数。C++程序的执行是从main()函数开始的，最初的函数调用也是由main()函数发起的，函数之间在经过不同的调用之后最后再返回main()函数结束程序。

函数嵌套调用的形式如下：

```
myFun1(){
  …
  myFun2();
  …
}


myFun2(){
  …
}


main(){
  …
  myFun1();
  …
}
```

程序开始执行时，必须先执行main()函数，然后由main()函数调用自定义函数myFun1()，后者再调用自定义函数myFun2()。

对于函数的嵌套调用需要注意以下两点：

（1）虽然C++允许函数嵌套调用，但是不允许函数嵌套定义，也就是说不能在定义函数时又包含另外的函数。

（2）函数嵌套调用时要注意不能出现死循环。

```
myFun1(){
  …
  myFun2();
  …
}
```

```
myFun2(){
  …
  myFun1();
}


main(){
  …
  myFun1();
  …
}
```

函数myFun1()在调用函数myFun2()，而函数myFun2()又返回来调用函数myFun1()，这种情况下很容易出现死循环。

函数之间在进行互相调用时，原则上是不允许调用main()函数的，只能在各个自定义函数之间互相调用。

下面这个例子演示了函数的嵌套调用。

实例4-14 用嵌套法做四则运算。

```
#include <iostream.h>


void inputNum();
void add(short int a, short int b);
void sub(short int a, short int b );
void mul(short int a, short int b );
void div(short int a, short int b );
short int check(short int m);


/* 输入数值函数 */
void inputNum(){
    short a,b;

    cout<<"a=";
    cin>>a;
    cout<<"b=";
    cin>>b;


    add(a,b);
```

```
}

/* 加法函数 */
void add(short int a, short int b){
    long result;

    result=a+b;
    cout<<a<<"+"<<b<<"="<<result<<"\n";
    sub(a,b);
}

/* 减法函数 */
void sub( short int a, short int b ){
    short int result;

    result=a-b;
    cout<<a<<"-"<<b<<"="<<result<<"\n";
    mul(a,b);
}

/* 乘法函数 */
void mul(short int a, short int b){
    long result;

    result=a*b;
    cout<<a<<"*"<<b<<"="<<result<<"\n";
    div(a,b);
}

/* 除法函数 */
void div(short int a, short int b){
    long result;

    if (check(b)==0)  {
        result=a/b;
        cout<<a<<"/"<<b<<"="<<result<<"\n";
```

```
    }
    else
        cout<<"除数不能为 0!"<<"\n";
}


/* 检查除数是否为 0*/
short int check(short int m){
    if (m==0)
        return 1;
    else
        return 0;
}


main(){
    cout<<"**************************"<<"\n";
    cout<<"* 用函数嵌套做加减乘法 *"<<"\n";
    cout<<"**************************"<<"\n";
    inputNum();
    cout<<"**************************"<<"\n";
}
```

程序运行结果如下：

```
**************************
* 用函数嵌套做加减乘法 *
**************************
a=1
b=2
1+2=3
1-2=-1
1*2=2
1/2=0
**************************
```

如果输入的除数为0，则程序运行结果如下：

```
**************************
* 用函数嵌套做加减乘法 *
**************************
a=1
```

```
b=0

1+0=1

1-0=1

1*0=0

除数不能为0！

************************
```

本例中定义了6个自定义函数inputNum()、add(short int a, short int b)、sub(short int a, short int b)、mul(short int a, short int b)、div(short int a, short int b)和short int check(short int m)。这6个函数分别用来实现不同的功能：

* 函数inputNum 要求用户输入数值，并将用户输入的数值作为实参传递给函数add。
* 函数add 利用形参接收函数inputNum传递过来的数值，并做加法运算，然后输出计算结果，并调用函数sub。
* 函数sub 利用形参接收函数add传递过来的数值，并做减法运算，然后输出计算结果，并调用函数mul。
* 函数mul 利用形参接收函数sub传递过来的数值，并做乘法运算，然后输出计算结果，并调用函数div。
* 函数div 利用形参接收函数mul传递过来的数值，然后调用函数check来判断接收的参数是否为0，如果为0则给出输出提示"除数不能为0！"，并不做任何运算；如果参数不为0则做除法运算，然后输出计算结果。

这个程序由main()函数开始执行，然后调用inputNum函数，后者再调用add函数，依次类推，调用关系如图4-1所示。

4.9 内联函数

在C++程序中，有些小函数在程序中被频繁调用，这样会大量消耗栈空间。所谓栈空间就是存放函数内的数据的内存空间。因为栈空间是有限的，如果频繁、大量地使用这些空间，就会造成堆栈空间不足。事实上，系统在很多情况下都是因为栈空间不足而导致瘫痪，例如前面讲C++的递归调用时，曾经提到过函数的死循环，这种调用的后果就是系统内存栈空间被耗尽。

函数的调用还会带来程序执行效率问题。函数调用会打乱程序执行的顺序，当一个函数调用另外一个函数时，需要先保护现在程序执行处的现场并保存执行的地址；然后将执行顺序转移到被调用的函数处，开始执行被调用函数；待被调用函数执行完毕后再返回调用函数处，然后恢复现场，并按照保存的地址继续执行。因此，函数调用会产生一些时间和空间上的系统开销，这将影响程序的执行效率。

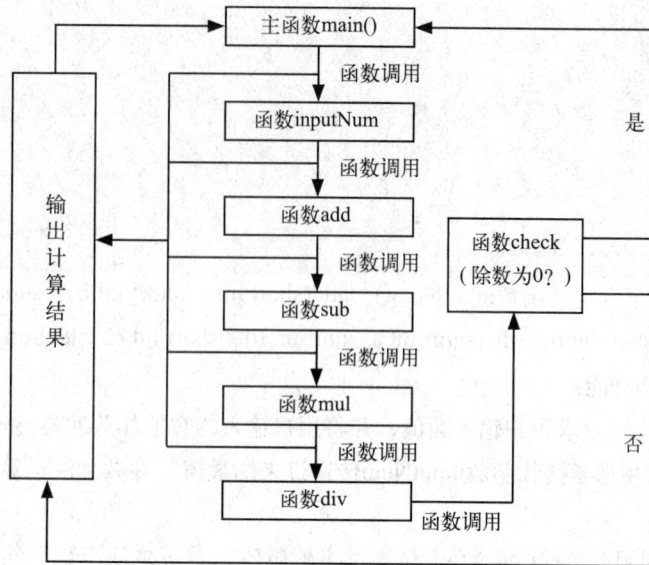

图4-1　函数嵌套调用

为了解决这两个问题，C++引入了内联函数的概念。内联函数就是由inline修饰符修饰的函数。

C++引入内联函数的目的在于，解决函数频繁调用导致系统内存栈被大量占用而出现系统内存栈枯竭的问题和程序中函数调用效率问题。

在程序编译时，编译器将程序中调用内联函数的地方全部用内联函数的函数体代替，这就解决了函数在调用时要不断转换地址、保存现场和恢复现场的问题。

尽管内联函数可以解决上述函数调用时带来的程序执行效率和内存栈空间开销问题，但同时也带来了新的问题：由于程序在编译时编译器会把内联函数的函数体代码插入到调用内联函数的地方，因此会使整个程序的代码量增加，从而造成一定量的空间开销。因此，可以说程序中使用内联函数是以增加程序代码量的方式来使函数执行效率提高的。

内联函数的定义方法如下：

```
inline short int myFun2(short int a,short int b) {
    return(a+b);
}
```

可见，内联函数的定义很简单，只是在函数定义前增加了inline修饰符。程序在编译时将调用该内联函数的地方用该函数的函数体来代替，例如：

```
void myFun1(){
    …
    myFun2(x,y);
    …
}

inline short int myFun2(short int a,short int b) {
```

```
    return(a+b);
}


main(){
  …

  myFun1();

  …

}
```

上面这个程序在编译时，编译器会把函数myFun1中调用内联函数myFun2的语句myFun2(x,y)用内联函数的代码替换，以达到节省函数调用所产生的时间开销。

使用内联函数时必须注意以下几点：

（1）内联函数在第一次被调用之前必须先定义。

（2）在内联函数体内，不允许出现循环语句和开头语句。

（3）内联函数的代码要尽量少，以减少程序的整个代码量。

下面这个例子演示了内联函数的定义和调用方法。

实例4-15　内联函数的定义和调用。

```
#include <iostream.h>

void myFun1();
inline char myFun2(int a);

void myFun1(){
    short int x;

    for (x=0;x<=10;x++)
      cout<<x<<myFun2(x)<<"5"<<"\n";
}


char myFun2(int a){
    if (a<5)
      return'<';
    else if (a>5)
      return'>';
    else if (a==5)
      return'=';
}
```

```
main(){
    cout<<"*************************"<<"\n";
    cout<<"*        内联函数的调用        *"<<"\n";
    cout<<"*************************"<<"\n";
    myFun1();
    cout<<"*************************"<<"\n";
}
```

程序运行结果如下：

```
*************************
*        内联函数的调用        *
*************************
0<5
1<5
2<5
3<5
4<5
5=5
6>5
7>5
8>5
9>5
10>5
*************************
```

在本例中定义了一个内联函数myFun2，其作用是判断调用它的函数传递给它的参数是否大于5，如果大于5，则返回一个>符号；如果小于5，则返回一个<符号；如果等于5，则返回一个=符号。内联函数在这个程序中的作用通过运算结果却看不出来，事实上这个程序在编译阶段就把myFun2的函数体替换到调用它的函数myFun1中的调用语句处了。

4.10 函数重载

前面已经学习了函数的定义和使用方法。在实际开发中，还会遇到一类问题：需要进行一组极其类似的运算，但是参加这些运算的参数的数据类型却不一样，该如何处理呢？答案很简单：定义不同的函数即可。但是这样却带来了另外一个问题：从方便函数管理的角度来看，会希望这类函数具有相同的名字，这又如何解决呢？

C++引入了函数重载的概念来解决上面的问题。所谓函数重载就是一组拥有相同函数名，但是函数参数却不相同的函数。参数不同是指参数的数据类型、个数等不相同。例如下面这组函数的定义：

```
short int add(short int a,short int b);
short int add(short int a,short int b,short int c);
short int add(float a,float b);
short int add(float a,float b,float c);
```

上面4个函数的名字都叫add，但是它们的参数类型、数量却各不相同。函数重载要求编译器能够根据函数参数的类型、数量来确定应该调用哪个函数。

下面来看一个函数参数类型不同的函数重载的例子。

实例4-16 参数类型不同的函数重载。

```
#include <iostream.h>

inline  void funStar();
void inputNum();
int add(short int a, short int b);
float add(float a, float b);

void funStar(){
    cout<<"        ************        "<<"\n";
}

void inputNum(){
    short int a,b;
    float x,y;

    funStar();
    cout<<" 请输入整型值 :"<<"\n";
    cout<<"a=";
    cin>>a;
    cout<<"b=";
    cin>>b;
    cout<<"a+b="<<add(a,b)<<"\n";
    funStar();
    cout<<" 请输入浮点型值 :"<<"\n";
    cout<<"x=";
    cin>>x;
```

```
        cout<<"y=";

        cin>>y;

        cout<<"x+y="<<add(x,y)<<"\n";

        funStar();

}


int add(short int a, short int b){

    return(a+b);

}


float add(float a, float b){

     return(a+b);

}


main(){

    cout<<"***************************"<<"\n";

    cout<<"*    参数不同的函数重载    *"<<"\n";

    cout<<"***************************"<<"\n";

    inputNum();

    cout<<"***************************"<<"\n";

}
```

程序运行结果如下：

```
***************************

*    参数不同的函数重载    *

***************************

          *************

请输入整型值：

a=1

b=2

a+b=3

          *************

请输入浮点型值：

x=1.2

y=2.3

x+y=3.5

          *************
```

这个例子重载了函数add。函数add有两个实现方式：第一个是参数为short int类型的实现方式，第二个是参数为float类型的实现方式。当输入两个short int类型的值时，程序会自动调用add函数的第一个实现方式；当输入两个float类型的值时，程序会自动调用add函数的第二个实现方式。在这个例子中还定义了一个内联函数来输出星号，以区别两个函数的输出显示结果。

下面再来看一个参数数量不同的函数重载的例子。

实例4-17 参数数量不同的函数的重载。

```cpp
#include <iostream.h>

inline void funStar();
void inputNum();
int add(int a, int b);
int add(int a, int b, int c);

void funStar(){
    cout<<"        *************        "<<"\n";
}

void inputNum(){
    int a,b,c;

    funStar();
    cout<<" 请输入 2 个整型值："<<"\n";
    cout<<"a=";
    cin>>a;
    cout<<"b=";
    cin>>b;
    cout<<"a+b="<<add(a,b)<<"\n";
    funStar();
    cout<<" 请输入 3 个整型值："<<"\n";
    cout<<"a=";
    cin>>a;
    cout<<"b=";
    cin>>b;
    cout<<"c=";
    cin>>c;
```

```
        cout<<"a+b+c="<<add(a,b,c)<<"\n";
        funStar();
}


int add(int a, int b){
        return(a+b);
}


int add(int a, int b, int c){
        return(a+b+c);
}


main(){
    cout<<"****************************"<<"\n";
    cout<<"*    参数不同的函数重载    *"<<"\n";
    cout<<"****************************"<<"\n";
    inputNum();
    cout<<"****************************"<<"\n";
}
```

程序运行结果如下：

```
****************************
*    参数不同的函数重载    *
****************************
        *************
请输入 2 个整型值：
a=1
b=2
a+b=3
        *************
请输入 3 个整型值：
a=1
b=2
c=3
a+b+c=6
        *************
****************************
```

在上例中重载了参数数量不同的函数add，这两个函数分别实现对两个数相加和三个数相加的操作。

函数重载在C++中是一个非常重要的概念，我们在后面的章节中还会讲到类和对象的概念，函数重载在类和对象的应用中比较多，尤其是在类的多态性应用方面。

4.11　上机操作

调用两个函数，分别求1～20之内的奇数及偶数。

实例4-18　求1～20之内的奇数及偶数。

```
#include <iostream.h>

void funEven();
void funOdd();

void funEven(){
    short int i;

    for (i=1;i<=20;i++)  {
        if (i%2==0)  {
            cout<<i<<" ";
        }
    }
    cout<<"\n";
}

void funOdd(){
    short int i;

    for (i=1; i<=20; i++)  {
        if (i%2!=0)  {
            cout<<i<<" ";
        }
    }
    cout<<"\n";
}
```

```
main(){
    cout<<"*************************************"<<"\n";
    cout<<"*        求1～20之间的奇偶数        *"<<"\n";
    cout<<"*************************************"<<"\n";
    cout<<"*              偶数如下              *"<<"\n";
    funEven();
    cout<<"*              奇数如下              *"<<"\n";
    funOdd();
    cout<<"*************************************"<<"\n";
}
```

程序运行结果如下：

```
*************************************
*        求1～20之间的奇偶数        *
*************************************
*              偶数如下              *
2  4  6  8  10  12  14  16  18  20
*              奇数如下              *
1  3  5  7  9  11  13  15  17  19
*************************************
```

使用循环求斐波那契数列。斐波那契数列是由意大利中世纪数学家列昂纳多·斐波那契发明的。指的是这样一列数列：1, 1, 2, 3, 5, 8, 13, 21, 34, 55, 89, 144, 233, …这个数列的前两项都是1，从第三项起，每一项都等于前两项的和。

斐波那契数列的通项公式可以写成下面的形式：

$$f(1)=f(2)=1, f(n)=f(n-1)+f(n-2) \quad (n>=3)。$$

下面是一个用循环来求斐波那契数列的n项值的程序。

实例4-19   用for循环求斐波那契数列的n项值。

```
/* 循环法求斐波那契数列 */
#include <iostream.h>

void inputNum();
void funFib(short int n);

void inputNum(){
    short int m;

    cout<<"m=";
```

```
    cin>>m;
    funFib(m);
}

void funFib(short int n){
    short int m;
    int a,b,sum;
    a=0;
    b=1;

    cout<<"1"<<"";
    for (m=1;m<n;m++)  {
        sum=a+b;
        a=b;
        b=sum;
        cout<<sum<<"";
    }
    cout<<"\n";
}

main(){
    cout<<"*****************************"<<"\n";
    cout<<"*    循环法求斐波那契数列    *"<<"\n";
    cout<<"*****************************"<<"\n";
    inputNum();
    cout<<"*****************************"<<"\n";
}
```

程序运行结果如下:

```
*****************************
*    循环法求斐波那契数列    *
*****************************
m=10
1 1 2 3 5 8 13 21 34 55
*****************************
```

在这个例子中给出了要求斐波那契数列前10项的值,程序运行后输出前10项的值。

使用递归法求斐波那契数列。

实例4-20 用递归法求斐波那契数列的n项值。

```cpp
/* 递归法求斐波那契数列 */
#include <iostream.h>

void inputNum();
int funFib(short int n);

void inputNum(){
    short int a;
    short int x;

    cout<<"x=";
    cin>>x;

    for (a=1;a<=x;a++)
        cout<<funFib(a)<<"";
    cout<<"\n";
}

int funFib(short int n){
    if(n==0)
        return 0;
    else if(n==1||n==2)
        return 1;
    else
        return funFib(n-1)+funFib(n-2);
}

main(){
    cout<<"*****************************"<<"\n";
    cout<<"*      递归法求斐波那契数列      *"<<"\n";
    cout<<"*****************************"<<"\n";
    inputNum();
    cout<<"*****************************"<<"\n";
}
```

程序运行结果如下：

```
*****************************
*     循环法求斐波那契数列     *
*****************************
x=10
1 1 2 3 5 8 13 21 34 55
*****************************
```

关于斐波那契数列还有许多算法，请读者依据此前学的知识，多多开动脑筋，研究还有什么算法，并一定亲自上机实验。

实例4-21 利用内联函数来判断数值的奇偶性。

```cpp
#include <iostream.h>

inline bool funEvenOrOdd(int n);
short int myFun();

/* 布尔类型的内联函数，用来判断数字的奇偶性 */
inline bool funEvenOrOdd(int n){
    return(n%2==0?0:1);
}

short int myFun(){
    short int x;
    int m;

    cout<<"x=";
    cin>>x;

    for (m=1;m<=x;m++)  {
        if (funEvenOrOdd(m))
            cout<<" 数字 : "<<m<<" 是奇数 "<<"\n";
        else
            cout<<" 数字 : "<<m<<" 是偶数 "<<"\n";
    }
}

main(){
    cout<<"*****************************"<<"\n";
```

```
    cout<<"*   利用内联函数判断数字的奇偶性  *"<<"\n";

    cout<<"********************************"<<"\n";

    myFun();

    cout<<"********************************"<<"\n";

}
```

程序运行结果如下：

```
********************************

*   利用内联函数判断数字的奇偶性  *

********************************

x=10

数字：1  是奇数

数字：2  是偶数

数字：3  是奇数

数字：4  是偶数

数字：5  是奇数

数字：6  是偶数

数字：7  是奇数

数字：8  是偶数

数字：9  是奇数

数字：10  是偶数

********************************
```

实例4-22   利用函数重载来比较数值大小。

```
#include <iostream.h>

void inputNum();
int max(int a, int b);
int max(int a, int b, int c);
int max(int a, int b, int c, int d);

void inputNum(){
    int a,b,c,d;

    cout<<"a=";
    cin>>a;
    cout<<"b=";
    cin>>b;
```

```
    cout<<"c=";
    cin>>c;
    cout<<"d=";
    cin>>d;

    cout<<"******* 比较结果 *********"<<"\n";
    cout<<"a 与 b 中较大者 : "<<max(a,b)<<"\n";
    cout<<"a、b 与 c 中较大者 : "<<max(a,b,c)<<"\n";
    cout<<"a、b、c 与 d 中较大者 : "<<max(a,b,c,d)<<"\n";
}

int max(int a, int b){
    if (a>b)
        return (a);
    else
        return (b);
}

int max(int a, int b, int c){
    int m;

    m=max(a,b);
    return max(m,c);
}

int max(int a, int b, int c, int d){
    int m1;
    int m2;
    m1=max(a,b);
    m2=max(c,d);
    return max(m1,m2);
}

main(){
    cout<<"***********************************"<<"\n";
    cout<<"*      利用函数重载比较数字大小      *"<<"\n";
```

```
        cout<<"*********************************"<<"\n";

        inputNum();

        cout<<"*********************************"<<"\n";

}
```

程序运行结果如下:

```
*********************************

*     利用函数重载比较数字大小     *

*********************************

a=26359

b=26401

c=30923

d=32221

******* 比较结果 *********

a 与 b 中较大者：26401

a、b 与 c 中较大者：30923

a、b、c 与 d 中较大者：32221

*********************************
```

在实际应用中，经常会遇到不同数值间比较大小的问题，而用户在实际应用中却往往不愿关心比较的具体实现方法。在本例中利用函数重载便实现了让用户不必关心数值比较大小的过程，只须调用相同名称的函数max，根据需要给出不同数量的参数即可。

例如在本例中，要比较两个数值的大小，程序调用函数max(int a,int b)，要比较3个数值的大小，则程序调用函数max(int a,int b,int c)，要比较4个数值的大小，程序调用函数max(int a,int b,int c,int d)即可。

4.12　小结

本章讲述了C++中的函数、函数调用、函数重载、内联函数、全局变量、局部变量的概念和使用方法。函数在程序设计语言中占有重要的地位，一个程序要实现很多功能，必然要把这些功能划分成若干小功能，而每个函数都是一个小功能的实现体。本章结合大量的实例来讲解函数的定义和实现方法。

在C++中，所有变量都具有作用域。据此，可以把变量划分成局部变量和全局变量。综合使用这两种变量，可以实现程序的不同功能。

4.13　习题

一、填空题

1. _____函数是一个特殊的函数，它是C++程序中唯一可以直接执行的函数，其他函数都是直接或间接被调用来执行的。

2．在定义具有参数的函数时，放在函数名后面的括号中的变量被称为_____，简称_____。

3．当有函数要调用有参函数时，必须把与_____中变量的____、变量的____、变量的____相同的变量传递给它才可以调用，这些变量被称为_____，简称____。

4．C++中的变量都有其作用域，根据变量定义的位置不同，其作用域也不同。据此，可以将C++中的变量分为_____和_____。

5．函数的递归调用是指一个函数在它的函数体内，直接或间接地调用_____。

二、程序阅读题

下例是用海伦公式求三角形面积的程序。海伦公式是这样描述的：

```
S=sqrt(1*(1-a)*(1-b)*(1-c));
```

其中，a、b、c 分别代表三角形的三条边，公式中 l=(a+b+c)/2。

要使三角形成立，还需具备这样的必要条件：三角形任意两条边长度之和必须大于第三边；三角形任意两条边长度之差必须小于第三边。

请补充完整下面的程序，使程序执行后输出如下的结果。

（1）假设两边长之和大于第三边，则输出如下结果：

```
******************************

*    用海伦公式求三角形的面积    *

******************************

请分别输入三角形三边的长：

a=3

b=4

c=5

该三角形的面积为：6

******************************
```

（2）假设两边长之和小于或等于第三边，则输出如下结果：

```
******************************

*    用海伦公式求三角形的面积    *

******************************

请分别输入三角形三边的长：

a=1

b=2

c=3

a、b、c 三条边不能构成三角形！

******************************
```

程序如下：

```
#include <iostream.h>
```

```
#include <math.h>

void inputNum();
bool funLine(float x,float y,float z);
float funArea(float a,float b,float c);

void inputNum(){
    float a,b,c;

    cout<<" 请分别输入三角形三边的长 :"<<"\n";
    cout<<"a=";
    cin>>a;
    cout<<"b=";
    cin>>b;
    cout<<"c=";
    cin>>c;
    if (_____)
        cout<<"a、b、c 三条边不能构成三角形 !"<<"\n";
    else
        cout<<" 该三角形的面积为 :"<<_____<<"\n";
}

/* 三角形成立判断 */
bool funLine(float x,float y,float z){
    if (_____)
        return 0;
    else
        return 1;
}

/* 求三角形的面积 */
float funArea(float a,float b,float c){
    float l,s;

    l=_____
    s=_____
```

```
        return(s);
    }


main(){
    cout<<"*******************************"<<"\n";
    cout<<"*    用海伦公式求三角形的面积    *"<<"\n";
    cout<<"*******************************"<<"\n";
    inputNum();
    cout<<"*******************************"<<"\n";
}
```

在这个示例中定义了一个函数funLine，用来判断用户输入的三角形的三条边能否构成三角形，这个函数的返回值应该是一个bool类型的值；然后函数inputNum再根据函数funLine的返回值来确定三角形是否成立，如果成立再调用函数funArea计算三角形的面积，否则给出提示"a、b、c三条边不能构成三角形!"。

三、问答题

1．函数在形式上可以分为几种？分别是什么函数？

2．有参函数和无参函数有什么不同？

3．什么是局部变量？什么是全局变量？请分别举例说明。

4．C++中定义的变量的存储类型有几种？分别是什么？

5．什么是函数的递归调用？使用递归调用时需要注意什么事项？

6．什么是函数的嵌套调用？

第5章　存储类型、生命周期和头文件

本章学习目标

➘ 外部存储类型　　➘ 静态存储类型
➘ 作用域与可见性　➘ 头文件

变量的存储类型可分为外部存储类型和静态存储类型两种。本章将对每种存储类型进行实例讲解。所有变量都有它的存在时间，称为变量的生命周期。头文件在C++程序中被大量使用，头文件是一种包含功能函数、数据接口声明的文件。在C++中，头文件以.h作为文件后缀。

5.1　外部存储类型

C++的变量存储类型可以分为外部存储类型和静态存储类型两种，本节将讲解外部存储类型，第5.2节讲解静态存储类型。

在以前的例子中，变量定义都是放在一个程序文件中，并且在这个文件中使用这些变量。那么，如果需要在另一个程序文件中也使用这个文件中定义的变量，该如何处理呢？这就是本节要讲的变量的外部存储类型。外部存储类型的概念不单单适用于变量，也适用于函数，即在一个程序文件中声明的函数，也可以在另外的程序文件中使用，这与变量是同样的道理。

外部存储类型的说明符是extern。下面这个例子将在一个程序文件中定义两个全局变量，而在另外一个程序文件中使用extern将这两个变量声明为外部变量，这样就可以在第二个程序文件中使用这两个变量了。例如：

```
//pro1.cpp
int x,y;
main(){
    …
}
//pro2.cpp
extern x,y;
int myFun(){
    …
}
```

在上面的例子中，有两个C++源文件pro1.cpp和pro2.cpp，在pro1.cpp中定义了两个全局变量x和y，在pro2.cpp中要使用这两个变量，需要使用extern把x和y说明为外部变量，这样等于告诉编译器这两个变量是在别的文件中声明的，不必再为它们分配空间，这样在pro2.cpp中就可以使用这两个变量了。

对于函数也是同样的道理。

```
//pro1.cpp
void exFun();
void exFun(){
    …
}
main(){
    …
}
//pro2.cpp
extern exFun();
int myFun(){
    …
}
```

5.2　静态存储类型

静态存储类型用说明符static声明。被static标识的全局变量和函数都被限制只能在当前源文件中使用。静态存储类型的定义格式如下：

```
static 变量或函数定义；
```

例如：

```
static int x;

static float y;
static int myFun();
```

下面这个例子在程序文件中定义了两个全局变量，这两个变量只能在该程序文件中使用，而不能被其他程序文件使用。

```
//pro.cpp
static int x;
static float y;
main(){
    …
}
```

在这个例子中，全局变量x和y被定义成static类型的变量，这样它们就只能在当前程序文件中使用，而不能被其他程序文件使用。如果其他文件要使用这两个变量，则必须使用extern来说明变量，而不能使用static。

对于静态函数也是同样的道理。

```
//pro.cpp
static void myFun1();
static int myFun2();


void myFun1(){
    …
}
void myFun2(){
    …

}


main(){
    …

}
```

由于静态存储类型限制了变量或函数只能在当前程序文件中使用，因此在有多个程序文件的工程中，可以在不同的文件中定义同名变量和函数，它们互不干扰，完全可以独立应用。例如，在一个工程文件中包含3个C++源文件，分别如下：

```
//pro1.cpp
static float age;
…
//pro2.cpp
static int age;
…
//pro3.cpp
static char age;
…
```

在这3个程序文件中都定义了变量age，但是它们在不同的程序文件中具有不同的类型。事实上，这3个变量就是3个完全不同的变量。

5.3 作用域与可见性

在前面讲变量定义的时候曾讲过，在同一个作用范围内，变量不能重复定义，例如：

```
…
float x;
float y;
x=0.5;
y=0.6
float x;
…
```

在这段代码中定义了两个变量x和y。但是在为变量赋值后又定义了一次变量x，在编译时就会出错。

通常，不同的作用域可以用一对花括号来区分，例如：

```
//pro.cpp
int myFun1(){
  float x;
  float y;
  …
}
float myFun2(){
  float a;
  int b;
  …
}
```

这个例子中包含两个函数，而且在每个函数中都声明有自己的变量，这些变量只在自己的函数体内（两个花括号之间）有效，出了这个范围这些变量便不存在。

另外一种情况如下：

```
//pro.cpp
float myFun(){
  int i;
  float x;
  …
  cin>>i;
  if (i>0)  {
    float x;
    …
  }
  x=100.5;
}
```

这个例子在函数中定义了一个float型变量x，在if程序块中也定义了一个同名变量x。执行时，if程序块中的变量x只在if块中有效，出了这个程序块就会是第一次定义的变量x生效，这称为可见性。对于函数也是同样的道理。

5.4 头文件

在C++中，可以把公用函数、结构等公用信息写在一个后缀名为.h的文件中，这种文件称为头文件。在C++源文件中，只要使用#include命令把相应的头文件包含进来，就可以使用其中定义的公用资源了。

例如，定义一个头文件arithmetic.h，它的作用是实现简单的四则运算。

```
/* 加法 */
float add(float x, float y){
    float result;

    result = x+y;
    return result;
}


/* 减法 */
float sub(float x, float y){
    float result;

    result = x-y;
    return result;
}


/* 乘法 */
float mul(float x, float y){
    float result;

    result = x*y;
    return result;
}


/* 除法 */
float div(float x, float y){
```

```
    float result;


    result = x/y;

    return result;

}
```

然后在源文件中编写调用它的代码：

```
#include <iostream.h>

#include <arithmetic.h>


void calcFun();


void calcFun(){

    float x, y;


    cout<<"x=";

    cin>>x;

    cout<<"y=";

    cin>>y;


    cout<<"x+y="<<add(x,y)<<endl;

    cout<<"x-y="<<sub(x,y)<<endl;

    cout<<"x*y="<<mul(x,y)<<endl;

    cout<<"x/y="<<div(x,y)<<endl;

}

main(){

    cout<<"*****************************"<<endl;

    cout<<"*      自定义头文件的使用      *"<<endl;

    cout<<"*****************************"<<endl;

    calcFun();

    cout<<"*****************************"<<endl;

}
```

把这个头文件放在公用目录下，以后编写程序时，只要将这个头文件包含进来就可以使用其中的函数了。

程序运行结果如下：

```
*****************************

*      自定义头文件的使用      *
```

```
****************************
x=90.9
y=100.5
x+y=191.4
x-y=-9.6
x*y=9135.45
x/y=0.904478
****************************
```

C++本身自带了非常丰富的头文件以供程序员调用，例如在前面例子中经常用到的iostream.h就是其中之一。

5.5 小结

本章讲述了C++中的存储类型（外部存储类型、静态存储类型）、生命期和头文件的知识。C++中的变量根据其在不同程序文件中的用途不同，可以设置成不同的存储类型，所有变量都有其生命期。头文件中存放的是一些公用信息，例如公用函数、结构等，这些定义放在头文件中方便在程序文件中调用。

5.6 习题

一、填空题

1. 变量的存储类型可分为_____存储类型和_____存储类型两种。

2. 外部存储类型的说明符是：_____；静态存储类型的说明符是：_____。

二、程序题

1. 编写一个两个数值比较大小的头文件，并在程序中调用这个头文件，完成对数值大小的比较。

2. 编写一个计算三角形面积、周长的头文件，并在程序中调用这个头文件，完成对三角形面积、周长的计算。

第6章 数 组

本章学习目标

➤ 数组的概念　　➤ 数组的定义和初始化

➤ 数组的访问方式　➤ 字符串

➤ 常用字符串处理函数

数组把具有相同类型的若干变量按照一定的顺序组织起来，这些按序排列的同类型数据元素的集合称为数组。组成数组的元素称为数组元素，按照数组元素的类型不同，可以把数组分成数值数组、字符数组、指针数组等。

6.1 数组的定义

数组是C++的一种数据类型，它的使用遵循C++变量的使用规则，即先定义后使用。数组定义的一般形式如下：

```
类型说明符 数组名 [ 常量表达式 ], …;
```

其中，类型说明符是任意一种基本数据类型或构造数据类型。数组名是用户定义的数组标识符。方括号中的常量表达式则表示数据元素的个数，也称为数组的长度。例如：

```
short int a[5];  //定义一个 short int 型的数组 a，该数组有 5 个元素
char b[20];      //定义一个 char 类型的数组 b，该数组有 20 个元素
```

对于数组的应用，应注意以下几点：

（1）数组是C++的一种数据类型，用它定义的变量同样遵循变量定义和使用规则。

（2）数组的类型实际上是数组中各元素的数据类型。

（3）数组定义时，在方括号中的常量表达式表示该数组元素的个数。

例如上面定义了一个数组short int a[5]，代表short int型数组a有5个元素。第一个元素从0开始计数，各数组元素分别用a[0]、a[1]…a[4]表示。假设数组a[5]中各元素的值依次是100、200、300、400、500，那么各数组元素可以分别表述如下：

```
a[0]=100

a[1]=200

a[2]=300

a[3]=400

a[4]=500
```

（4）方括号内的常量表达式不能用变量表示。

```
short int n;

short int a[n];
```

上面的定义是错误的，因为系统在为数组开辟内存空间时，必须知道需要开辟的内存单元的数量。下面的定义则是正确的：

```
#define NUM 2;

main(){

    int a[5];

    int b[3+2];

    int b[3+NUM];

}
```

6.2 访问数组元素

对数组元素的访问可以分为读数组元素和写数组元素。读、写数组元素都是通过数组下标变量来完成的。数组下标变量以[n]符号表示，其中字母n可以是数字，也可以是常量、常量表达式或变量。例如：

```
#define NUM 2;

main(){

    int a[6];

    a[3+NUM]=2;

    …

}
```

a[n]=1，其中n为变量。

以上都是访问数组元素的正确方法。

数组一旦被定义就可以非常方便地使用数组元素了，下面的例子演示了数组在被定义并被赋初值后是如何被读取出来的。

实例6-1 读取数组元素。

```
#include <iostream.h>

void myArray();

void myArray(){

    short int a[5]={1,2,3,4,5}; /* 定义数组 a 并为其赋初值 */

    short int m;

    cout<<"a[5]={1,2,3,4,5}"<<"\n\n";
```

```
        for (m=0;m<=4;m++)  {
            cout<<"a["<<m<<"]="<<a[m]<<"\n"; /* 根据下标变量读取数组元素的值 */
        }
}


main(){
    cout<<"************************"<<"\n";
    cout<<"*      数组元素的读取      *"<<"\n";
    cout<<"************************"<<"\n";
    myArray();
    cout<<"************************"<<"\n";
}
```

程序运行结果如下:

```
************************
*      数组元素的读取      *
************************
a[5]={1,2,3,4,5}

a[0]=1
a[1]=2
a[2]=3
a[3]=4
a[4]=5
************************
```

在这个例子中，首先定义了一个short int类型的数组a，定义数组包含5个数组元素，并且已给数组赋初值。

关于数组赋初值的问题，将在第6.3节做介绍。值得注意的是，C++访问数组元素是通过数组的下标变量来进行的，而C++本身并不对数组的下标变量值进行检查。可以这样理解：我们定义一个数组a，它有n个元素，那么这个数组下标变量值的有效范围就是0到n-1（因为下标变量值是从0开始的，而不是从1开始）。如果在读取数组元素时使用的数组下标变量值超出这个范围，会如何呢？我们将实例6-1稍做修改，将下面的循环语句:

```
for (m=0; m<=4; m++)  {
    cout<<"a["<<m<<"]="<<a[m]<<"\n"; /* 根据下标变量读取数组元素的值 */
}
```

修改成:

```
for (m=0; m<=5; m++)  {
```

```
                cout<<"a["<<m<<"]="<<a[m]<<"\n";  /* 根据下标变量读取数组元素的值 */
    }
```

因为定义的数组a只包含5个元素，并没有第6个元素，这时输出结果会是什么呢？实际运行程序后，结果如下：

```
*************************

*      数组元素的读取      *

*************************

a[5]={1,2,3,4,5}


a[0]=1

a[1]=2

a[2]=3

a[3]=4

a[4]=5

a[5]=34

*************************
```

显然，可以看到程序确实输出了第6个值，但是这个值令人感到莫名其妙，因为程序中不曾给数组赋予34这个值。这个问题就是使用数组时下标变量值越界的问题。

事实上，这个值是C++随机给出的值，因为第6个值在数组中是不存在的，但是C++并不检测下标变量值是否越界，这就需要程序员在定义和使用数组时，检查下标变量值是否越界。

对数组赋值也很简单，为一个数组赋值其实就是为数组中的每个数组元素逐个赋值。在为数组赋值时，也需要注意不要产生数组下标变量值越界的问题，以及所赋予的值的数据类型必须和数组元素的数据类型相匹配，否则会产生错误。

下面这个例子演示了如何为数组赋值。

实例6-2 为数组赋值。

```cpp
#include <iostream.h>


void myArray();


void myArray(){
    short int a[5];
    short int x,m;


    for (x=0;x<=4;x++)  {
        a[x]=(x+1)*(x-1);
    }
```

```
    for (m=0;m<=4;m++)  {
        cout<<"a["<<m<<"]="<<a[m]<<"\n";
    }
}


main(){
    cout<<"***********************"<<"\n";
    cout<<"*      数组元素的赋值      *"<<"\n";
    cout<<"***********************"<<"\n";
    myArray();
    cout<<"***********************"<<"\n";
}
```

程序运行结果如下：

```
***********************
*      数组元素的赋值      *
***********************
a[0]=-1
a[1]=0
a[2]=3
a[3]=8
a[4]=15
***********************
```

在这个例子中，首先通过一个for循环逐个地为数组的每个元素赋值，然后通过一个for循环逐个读出数组元素的值。将上面的程序稍做改动，就可以实现人机交互模式为数组赋值，将下面的语句：

```
for (x=0; x<=4; x++)  {
    a[x]=(x+1)*(x-1);
}
```

修改为：

```
for (x=0; x<=4; x++)  {
    cout<<" 请输入 a["<<x<<"] 的值 :";
    cin>>a[x];
}
```

修改后程序的运行结果则变成以下形式：

```
***********************
```

C++语言设计教程

```
*        数组元素的赋值        *
* * * * * * * * * * * * * * * * * * * * * * * *
请输入 a[0] 的值 :1
请输入 a[1] 的值 :2
请输入 a[2] 的值 :3
请输入 a[3] 的值 :4
请输入 a[4] 的值 :5
a[0]=1
a[1]=2
a[2]=3
a[3]=4
a[4]=5
* * * * * * * * * * * * * * * * * * * * * * * *
```

程序要求用户逐个输入数组中各个数组元素的值，然后再把用户输入的值原样读出来。人机交互界面大大增加了程序的可用性。

通过本节的学习可以了解到，读取数组的值实际上就是读取数组中各个数组元素的值，而为数组赋值实际上就是为数组的各个数组元素赋值。无论读取数组还是为数组赋值都需要注意以下两点：

（1）注意数组元素的下标变量值，不要越界。

（2）为数组元素赋值时，所赋予的值必须和数组元素的数据类型相匹配。

6.3 数组的初始化

在第6.2节中已经接触到数组初始化的一种方法，即：

类型说明符 数组名 [常量表达式]={ 值 1，值 2，…，值 n}；

例如int a[5]={1, 2, 3, 4, 5}，这条语句在定义时完成了该数组的初始化。赋值过程相当于以下5条语句：

```
a[0]=1;
a[1]=2;
a[2]=3;
a[3]=4;
a[4]=5;
```

数组除了在定义的同时进行初始化之外，还可以通过赋值语句进行初始化，而通过赋值语句初始化在实际应用中更为常见。

在实例6-2中已经演示了如何通过赋值语句为数组赋值。下面再看这样一个例子——用赋值语句初始化数组的示例。

实例6-3 通过赋值语句初始化数组。

164

```cpp
#include <iostream.h>
#include <math.h>

void myFun();

void myFun(){
    const short int NUM=5;

    short int m;
    double myArray[NUM];
    int x,y;

    x=1;
    y=2;

    for (m=0; m<=NUM-1; m++)  {
        if (m%2!=0)   {/* 为奇数数组元素赋值 */
            myArray[m]=pow(x,y);
        }  else  {
            myArray[m]=pow(y,x);  /* 为 0 和偶数数组元素赋值 */
        }
        x++;
        y++;
    }
    for (m=0; m<=NUM-1; m++)  {
        cout<<"myArray["<<m<<"]="<<myArray[m]<<"\n";
    }
}

main(){
    cout<<"********************************"<<"\n";
    cout<<"*        用赋值语句初始化数组        *"<<"\n";
    cout<<"********************************"<<"\n";
    myFun();
    cout<<"********************************"<<"\n";
}
```

程序运行结果如下：

```
******************************
*      用赋值语句初始化数组      *
******************************
myArray[0]=2
myArray[1]=8
myArray[2]=64
myArray[3]=1024
myArray[4]=7776
******************************
```

在这个程序中，首先通过一个for循环为数组myArray赋值。在for循环中，又使用了if条件判断语句来对数组元素的奇偶性进行判断，分别为奇数数组元素和偶数数组元素使用不同的算法进行赋值。

事实上，在为一个数组赋值时，可以不用为所有数组元素赋值，而只赋值一部分数组元素的值。将实例6-3中的for循环语句：

```
for (m=0; m<=NUM-1; m++)  {
    if (m%2!=0)  {/* 为奇数数组元素赋值 */
        myArray[m]=pow(x,y);
    } else {
        /* myArray[m]=pow(y,x); /* 为 0 和偶数数组元素赋值 */
    }
}
```

修改成如下形式：

```
for (m=0; m<=NUM-1; m++)  {
    if (m%2!=0)  {/* 为奇数数组元素赋值 */
        myArray[m]=pow(x,y);
    } else {
        /* 此处不为 0 和偶数数组元素赋初值 */
    }
}
```

程序运行结果如下：

```
******************************
*      用赋值语句初始化数组      *
******************************
myArray[0]=1.80106e-307
myArray[1]=8
```

```
myArray[2]=5.28312e-308

myArray[3]=1024

myArray[4]=2.25463e-307

*****************************
```

可以看到，数组元素的奇数项都按照程序中规定的算法为数组元素赋予了初值，但是其他项目却被赋予了一些奇怪的数字，这些数字是C++为数组中没有被赋值的元素随机分配的数字，它们没有任何意义。

6.4　一维数组的定义和使用

在前面介绍数组定义的时候已经讲过，数组元素是组成数组的基本单元。数组元素也是一种变量，其标识方法为数组名后跟一个下标。如果只有一个下标则是一维数组。下标表示元素在数组中的顺序号。数组元素的一般形式为：数组名[下标]，其中下标只能为整型常量或整型表达式。

本章前面所接触的数组均为一维数组。数组必须先定义，才能使用下标变量，在引用数组元素时，也是通过数组下标逐个引用的，而不能一次引用整个数组，例如：

```
char a[10];

cout<<a;
```

这种引用方法是错误的，只有通过下标变量引用，如：

```
for (short int i=0; i<=9; i++)

    cout<<a[i];
```

前面讲数组定义时讲过，数组定义的一般形式如下：

　类型说明符 数组名〔常量表达式〕，…；

如果在定义数组时就对它进行初始化，那么常量表达式可以省略，这样数组的定义和初始化可以变成以下形式：

　类型说明符 数组名[]={值1，值2，…，值n}；

即省略了常量表达式后，数组元素的个数由该数组初始化时的数组元素的个数来确定。

下面是一个利用数组排序的程序。

　实例6-4　利用数组排序。

```
#include <iostream.h>


void funSort();


void funSort(){
    int i,j,p,q,s;
    int a[]={97,100,1021,1122,991,987,925,1399,876,1199};
```

```
        cout<<" 排序结果如下 :"<<"\n";


        for(i=0;  i<10;  i++)   {
            p=i;
            q=a[i];
            for(j=i+1;  j<10;  j++)
                if(q<a[j])   {
                    p=j;
                    q=a[j];
                }
                if(i!=p)   {
                    s=a[i];
                    a[i]=a[p];
                    a[p]=s;
                }
                cout<<a[i]<<"";
        }


        cout<<"\n";
}


main(){
    cout<<"*********************************************"<<"\n";
    cout<<"*                利用数组排序                *"<<"\n";
    cout<<"*********************************************"<<"\n";
    funSort();
    cout<<"*********************************************"<<"\n";


}
```

程序运行结果如下:

```
*********************************************
*                利用数组排序                *
*********************************************
排序结果如下 :
1399 1199 1122 1021 991 987 925 876 100 97
*********************************************
```

　　在这个程序中定义了数组a，但并没有给出该数组的元素个数。在定义数组时就对其进行了初始化，为数组赋予了10个值，这样该数组就有10个元素。

　　本程序的排序采用逐个比较的方法进行。在i次循环时，把第一个元素的下标i赋予p，而把该下标变量值a[i]赋予q。之后进入小循环，从a[i+1]起到最后一个元素止逐个与a[i]做比较，凡是比a[i]大者则将其下标送回p，元素值送回q。

　　一次循环结束后，p即为最大元素的下标，q则为该元素值。若此时i≠p，说明p、q的值均已不是进入小循环之前所赋之值，则交换a[i]和a[p]之值。此时a[i]为已排序完毕的元素。输出该值之后转入下一次循环，再对i+1以后的各个元素排序。

　　在这个例子中，采用了数组定义时就初始化的方法，但这种方法的弊端也是显而易见的，那就是对要排序的数值不够灵活。假如需要再对另外一组10个数字进行排序，那么就必须修改这个数组的初始值，也就意味着需要修改程序。

　　为了解决这个问题，可以把数组的初始化方式更改为人机交互式，让用户为数组输入初始值。为了达到这个要求，需要对该程序进行两点修改：

　　（1）定义数组a时给出常量表达式的值。

　　（2）利用另外一个for循环语句逐个为数组a中的各元素赋值。

　　请读者根据上面两点的设想修改实例6-4的程序。

实例6-5　利用数组判断用户输入的值中的最大值和最小值。

```cpp
#include <iostream.h>

void funMaxAndMin();

void funMaxAndMin(){
    int m,min,max;
    int a[10];

    cout<<" 请输入你要排序的 10 个数字 :"<<"\n";
     for(m=0; m<10; m++)  {
         cout<<" 请输入第 "<<m+1<<" 个数字 :";
         cin>>a[m];
    }


     min=a[0];


for(m=1; m<10; m++)
        if(a[m]<min)
           min=a[m];
        else
```

```
                max=a[m];

        cout<<" 你输入的最小值是 :"<<min<<"\n\n";
        cout<<" 你输入的最大值是 :"<<max<<"\n";

}


main(){
    cout<<"*****************************"<<"\n";
    cout<<"*     利用数组判断最大最小值     *"<<"\n";
    cout<<"*****************************"<<"\n";
    funMaxAndMin();
    cout<<"*****************************"<<"\n";
}
```

程序运行结果如下：

```
*****************************

*     利用数组判断最大最小值     *

*****************************

请输入你要排序的 10 个数字 :
请输入第 1 个数字 :123
请输入第 2 个数字 :234
请输入第 3 个数字 :345
请输入第 4 个数字 :567
请输入第 5 个数字 :568
请输入第 6 个数字 :678
请输入第 7 个数字 :768
请输入第 8 个数字 :798
请输入第 9 个数字 :820
请输入第 10 个数字 :868


你输入的最小值是 :123
你输入的最大值是 :868

*****************************
```

本例中使用一个for循环语句来逐个接收用户输入的10个数到数组a中，然后把a[0]的值赋给变量min。在第二个for循环中，从a[1]到a[9]逐个与min中的内容做比较，若比min的值大，则把该下标变量的值赋给变量max，否则赋给变量min，因此变量max的值始终是最大值，而变量min的值始终是最小值。

6.5　二维数组的定义和使用

前面已经讲了一维数组，数组有一维数组和多维数组之分。一维数组只有一个下标变量，而多维数组的元素有多个下标，以标识它在数组中的位置。最简单的多维数组就是二维数组。实际上，二维数组就是以一维数组为元素构成的数组。定义二维数组的一般形式如下：

类型说明符　数组名 [常量表达式 1] [常量表达式 2] … ;

其中，常量表达式1表示该二维数组第一维下标的长度，常量表达式2则表示该二维数组第二维下标的长度。例如：

int a[3][4];

上面是一个名为a的整型二维数组，它由3行4列组成，其数组元素分别为：

a[0][0], a [0][1], a [0][2], a [0][3]

a[1][0], a [1][1], a [1][2], a [1][3]

a[2][0], a [2][1], a [2][2], a [2][3]

实例6-6为一个简单的二维数组赋初值，并将这个数组按顺序输出。

实例6-6　为二维数组赋值，并顺序输出二维数组的各元素值。

```
#include <iostream.h>
#include <math.h>

void funArray();

void funArray(){
    double myArray[3][4];
    int i,j;

    for (i=0; i<3; i++)   {/* 定义二维数组的行 */
        for (j=0;j<4;j++) /* 定义二维数组的列 */
            myArray[i][j]=pow(i,j);
    }

    for (i=0; i<3; i++) {
        for (j=0; j<4; j++)  {
            cout<<"myArray["<<i<<"]["<<j<<"]="<<myArray[i][j]<<"";
        }
        cout<<"\n";
    }
}
```

```
main(){
    cout<<"****************************************************************"<<"\n";
    cout<<"*                      二维数组的赋值和输出                      *"<<"\n";
    cout<<"****************************************************************"<<"\n";
    funArray();
    cout<<"****************************************************************"<<"\n";
}
```

程序运行结果如下：

```
****************************************************************
*                      二维数组的赋值和输出                      *
****************************************************************
myArray[0][0]=1 myArray[0][1]=0 myArray[0][2]=0 myArray[0][3]=0
myArray[1][0]=1 myArray[1][1]=1 myArray[1][2]=1 myArray[1][3]=1
myArray[2][0]=1 myArray[2][1]=2 myArray[2][2]=4 myArray[2][3]=8
****************************************************************
```

本例中定义了一个二维数组myArray[3][4]，它有3行4列。在程序中，首先用一个for循环嵌套分别为这个二维数组的每个元素赋值。赋值后，这个数组可以更直观地表现为图6-1所示的形式。

$$\left\{ \begin{matrix} 1 & 0 & 0 & 0 \\ 1 & 1 & 1 & 1 \\ 1 & 2 & 4 & 8 \end{matrix} \right\}$$

图6-1 二维数组

然后，再利用一个for循环嵌套把这个数组的值输出到屏幕上。

前面讲一维数组的时候曾讲到在定义一维数组的同时初始化数组，例如：

```
int a[3]={1,2,3};
```

二维数组同样可以这样做，例如：

```
int a[3][4]={1,0,0,0,1,1,1,1,1,2,4,8};
```

但是这样赋值并不直观，如果把这些值按照二维数组的行进行分组，在书写格式上稍做变化，能使赋值变得更直观、更容易理解。例如：

```
int a[3][4]={{1,0,0,0},{1,1,1,1},{1,2,4,8}};
```

按照书写顺序，每对大括号代表二维数组的一行，而大括号内的值则代表二维数组的一列。

在介绍一维数组时曾经讲过常量表达式可以省略，在二维数组中同样可以省略常量表达式，不过只能省略常量表达式1，例如：

```
int a[][4]={{1,0,0,0},{1,1,1,1},{1,2,4,8}};
```

这里省略了常量表达式1，系统会自动识别这个数组为一个3行4列的二维数组。

实例6-7 二维数组初始化时省略常量表达式。

```cpp
#include <iostream.h>


void funArray();


void funArray(){
    int a[][3]={{1,2,3},{4,5,6}};
    int i,j;


    for (i=0; i<2; i++)   {
        for (j=0;j<3;j++)
            cout<<"a["<<i<<"]["<<j<<"]="<<a[i][j]<<"";
        cout<<"\n";
    }
}


main(){
    cout<<"*****************************************"<<"\n";
    cout<<"        二维数组初始化时省略常量表达式 1        *"<<"\n";
    cout<<"*****************************************"<<"\n";
    funArray();
    cout<<"*****************************************"<<"\n";
}
```

程序运行结果如下：

```
*****************************************
        二维数组初始化时省略常量表达式 1        *
*****************************************
a[0][0]=1 a[0][1]=2 a[0][2]=3
a[1][0]=4 a[1][1]=5 a[1][2]=6
*****************************************
```

6.6 字符数组

用来存放字符变量的数组称为字符数组。定义字符数组的一般形式和前面讲到的一维数组、二维数组相同，例如：

```cpp
char a[3];
```

字符数组可以是一维数组、二维数组或多维数组，例如：

```
char a[3];

char a[3][4];
```

与前面讲到的一维数组、二维数组一样，也允许在定义字符数组的同时，对它进行初始化，例如：

```
char a[3]={'a','b','c'};

char a[]={'a','b','c','d','e','f'};

char a[2][3]={{'a','b','c'},{'d','e','f'}};

char a[][2]={{'a','b'},{'d','e' }};
```

实例6-8　字符数组初始化并输出数组元素值。

```
#include <iostream.h>


void funArray();


void funArray(){
    char a[][6]={{'s','y','b','a','s','e'},{'s','e','r','v','e','r'}};
    int i,j;


    for (i=0; i<2; i++)  {
        for (j=0;j<6;j++)
            cout<<"a["<<i<<"]["<<j<<"]="<<a[i][j]<<"";
        cout<<"\n";
    }
}


main(){
    cout<<"*****************************************************"<<"\n";
    cout<<"              字符数组初始化并输出数组元素值            *"<<"\n";
    cout<<"*****************************************************"<<"\n";
    funArray();
    cout<<"*****************************************************"<<"\n";
}
```

程序运行结果如下：

```
*****************************************************
          字符数组初始化并输出数组元素值              *
*****************************************************
```

```
a[0][0]=s a[0][1]=y a[0][2]=b a[0][3]=a a[0][4]=s a[0][5]=e

a[1][0]=s a[1][1]=e a[1][2]=r a[1][3]=v a[1][4]=e a[1][5]=r

***************************************************************
```

事实上，C++允许以字符串的形式来为字符数组赋值，形式如下：

```
char a[]={"sybase server"};
```

或者：

```
char a[]="sybase server";
```

这与逐个为字符数组的数组元素赋值是等效的。据此，可以把上面的例子修改为下面的实例6-9。

实例6-9　用字符串为字符数组赋值。

```
#include <iostream.h>


void funArray();


void funArray(){
    char a[]={"sybase server"};
    int i,j;


    for (i=0; i<13; i++)
        cout<<a[i];
    cout<<"\n";
}


main(){
    cout<<"************************************************************"<<"\n";
    cout<<"                字符数组整体初始化并输出数组元素值                *"<<"\n";
    cout<<"************************************************************"<<"\n";
    funArray();
    cout<<"************************************************************"<<"\n";
}
```

程序运行结果如下：

```
************************************************************
                字符数组整体初始化并输出数组元素值                *
************************************************************
sybase server
************************************************************
```

6.7 字符串

在C++中，字符串就是用一对双引号括起来的一串字符，如：

```
"abc"
"a+b+c"
"study c++"
"a+b=c\n"
"好好学习，天天向上！"
```

这些都是字符串。字符串的长度就是一对双引号之间的字符的个数（包括空格），需要注意的是，一个ASCII码的长度是一个字符，而一个区位码（比如汉字）的长度是两个字符。例如：

字符串"abc"的长度是3，而字符串"计算机"的长度是6。C++允许空字符串""，空字符串的长度是0。

C++中的字符串是以一维数组的形式来存储的，而在一维字符数组中，则是以字符\0作为该字符串的结束符。因此，存储字符串的一维字符数组的长度，必须大于字符串本身的长度。

例如：字符串"计算机教程"的长度是10，那么用来存储这个字符串的一维字符数组的长度至少应该是11。

字符串以一维字符数组的形式存放，可以用图6-2来表示。

```
char a[8]="sybase";
```

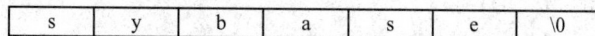

| s | y | b | a | s | e | \0 | |

图6-2 字符串在数组中的存放情况

在图6-2中，每个单元格代表一维字符数组的一个下标位，每个单元格都存放一个字符，最后以字符\0结束。实际上，这个一维字符数组剩下的下标位也是以字符\0来填充的，即最后那个单元格存放的也是字符\0。

在前面讲一维数组的时候就讲过，数组中各元素可以用数组的下标变量来表示，例如在上例中，字符串数组a的各元素可以表示如下：

```
a[0]='s';
a[1]='y';
a[2]='b';
a[3]='a';
a[4]='s';
a[5]='e';
a[6]='\0';
a[7]='\0';
```

对于字符串数组，也可以对它进行整体赋值和整体输出。

实例6-10 用一维数组来实现字符串的整体赋值与输出。

```
#include <iostream.h>
```

```
void funArray();

void funArray(){
    char a[8];

    cout<<"请输入字符串（最长 7 位）:";
    cin>>a;
    cout<<"你输入的字符串是:";
    cout<<a;
    cout<<"\n";
}


main(){
    cout<<"*****************************************************"<<"\n";
    cout<<"                    字符串的整体赋值与输出              *"<<"\n";
    cout<<"*****************************************************"<<"\n";
    funArray();
    cout<<"*****************************************************"<<"\n";
}
```

程序运行结果如下：

```
*****************************************************
              字符串的整体赋值与输出              *
*****************************************************
请输入字符串（最长 7 位）:sybase
你输入的字符串是:sybase
*****************************************************
```

本例中没有采用数组下标变量的方法来为字符数组赋值，也没有采用下标变量的办法来输出数组元素的值。而是直接将字符串赋予数组变量，并直接从数组变量读取数组的值。

需要注意的是，在使用这种方法来给字符数组赋值时，不允许在字符之间出现空格，否则C++会认为用户输入的字符已经结束，会自动结束这个字符串。像实例6-10那样，如果输入的字符串之间有空格，则程序运行结果如下：

```
*****************************************************
              字符串的整体赋值与输出              *
*****************************************************
请输入字符串（最长 7 位）:sybas e
```

你输入的字符串是:sybas

****

6.8 常用字符串处理函数

为了方便处理字符串，C++特别定义了一组处理字符串的函数，这些函数的原型被保存在名为string.h的头文件中。如果要使用这些函数的一种或多种，那么必须先使用#include语句将这个头文件包含进来，然后才可以在程序的任何地方使用这些函数。

> **提示**
>
> 本节将以实例讲解几个实用的字符串处理函数，这些函数在实际程序设计中会经常用到，请读者好好理解。

本节将学习C++中常用的几个字符串处理函数。

1. 测试字符串长度函数strlen

strlen函数用来测试字符串的长度，其函数原型如下：

```
int strlen(const char s[]);
```

由此函数的原型可以看出，该函数的返回值是int类型，代表字符串的长度。此函数仅有一个参数，这个参数可以是字符串，也可以是一个一维字符数组，还可以是二维数组中只带行下标的单下标变量。该函数的参数中有一个const关键字，表示该参数在程序中不能被改变。

实例6-11 使用strlen函数测试字符串长度。

```cpp
#include <iostream.h>
#include <string.h>


void inputString();
int funStrlen(char a[]);


void inputString(){
    char a[256];

    cout<<"请输入待测字符串:";
    cin>>a;
    cout<<"你输入的字符串长度为:"<<funStrlen(a);
    cout<<"\n";
}


int funStrlen(char a[]){
```

```
        return strlen(a);
}


main(){
    cout<<"*******************************"<<"\n";
    cout<<"*   测试字符串长度函数 -strlen    *"<<"\n";
    cout<<"*******************************"<<"\n";
    inputString();
    cout<<"*******************************"<<"\n";
}
```

程序运行结果如下：

```
*******************************
*   测试字符串长度函数 -strlen    *
*******************************
请输入待测字符串：sybase
你输入的字符串长度为：6
*******************************
```

本例中定义了两个函数：inputString函数用来接收用户输入的字符串，funStrlen函数实现用strlen函数测试字符串的长度。

值得注意的是，strlen函数测试的结果是字符串的实际长度，并不包含字符串最后的结束标识\0。

2. 字符串连接函数strcat

strcat函数用来连接两个字符串，其函数原型如下：

```
char *strcat(char *dest, const char *src);
```

其作用是把第二个参数src指向的字符串连接到第一个参数dest所指的字符串之后。这意味着第一个参数所指的字符串的存储空间的长度，必须大于或等于第一个参数和第二个参数所指的字符串的长度之和。

实例6-12　字符串连接函数strcat。

```
#include <iostream.h>
#include <string.h>


void inputString();


void inputString(){
    char a[20];
    char b[11];
```

```
    cout<<" 请输入第一个符串：";

    cin>>a;

    cout<<" 请输入第二个符串：";

    cin>>b;


    cout<<" 连接后的字符串为："<<strcat(a,b);

    cout<<"\n";

}


main(){

    cout<<"*********************************"<<"\n";

    cout<<"*      字符串连接函数 -strcat      *"<<"\n";

    cout<<"*********************************"<<"\n";

    inputString();

    cout<<"*********************************"<<"\n";

}
```

程序运行结果如下：

```
*********************************

*      字符串连接函数 -strcat      *

*********************************

请输入第一个符串：Hello

请输入第二个符串：World

连接后的字符串为：HelloWorld

*********************************
```

3．字符串拷贝函数strcpy

strcpy函数的原型如下：

```
char *strcpy(char *dest, const char *src);
```

strcpy函数的参数与返回值跟strcat函数相同，其调用格式为：

```
strcpy(字符数组 1，字符数组 2);
```

其作用是：把字符数组2中的字符串拷贝到字符数组1中，字符串结束标识\0也一同拷贝。字符数组2也可以是一个字符串常量，这时相当于把一个字符串赋予一个字符数组。在后面的章节里会学到数组指针的概念，有了这个概念后能对strcpy函数有更深的理解。

实例6-13　字符串拷贝函数strcpy。

```
#include <iostream.h>

#include <string.h>
```

```
void inputString();

void inputString(){
    char a[20]="";
    char b[11];

    cout<<"请输入符串：";
    cin>>b;
    cout<<"字符串 a 的值 ="<<a<<"\n";
    cout<<"字符串 b 的值 ="<<b<<"\n";

    cout<<"拷贝后字符串 a 的值 ="<<strcpy(a,b)<<"\n";
    cout<<"拷贝后字符串 b 的值 ="<<b<<"\n";
}

main(){
    cout<<"********************************"<<"\n";
    cout<<"*      字符串拷贝函数 -strcpy       *"<<"\n";
    cout<<"********************************"<<"\n";
    inputString();
    cout<<"********************************"<<"\n";
}
```

程序运行结果如下：

```
********************************
*      字符串拷贝函数 -strcpy       *
********************************
请输入符串：Hello
字符串 a 的值 =
字符串 b 的值 =Hello
拷贝后字符串 a 的值 =Hello
拷贝后字符串 b 的值 =Hello
********************************
```

从本例的运行结果可以看出，a 的值初始化为空字符串，b 的值为字符串"Hello"，在执行完字符串拷贝后，b 的值被拷贝到 a 中，这样 a 的值也就变成字符串"Hello"了。

和使用 strcat 函数一样，在使用 strcpy 时，要求接收方的数组一定要有足够的长度来接收被拷贝的字符串。

4．字符串比较函数strcmp

字符串比较函数用来比较两个字符串的大小，其函数原型如下：

```
int strcmp(const char *s1,const char *s2);
```

函数中两个指针参数分别指向两个字符串s1和s2，这两个字符串根据相比较后的大小返回不同的值：

- s1>s2 返回一个大于0的数。
- s1=s2 返回0。
- s1<s2 返回一个小于0的数。

实例6-14　两个字符串比较大小。

```cpp
#include <iostream.h>

void funArray();

void funArray(){
    char a[8];
    char b[8];

    cout<<"请输入字符串（最长 7 位）a=";
    cin>>a;
    cout<<"请输入字符串（最长 7 位）b=";
    cin>>b;
    cout<<"\n";

    if (strcmp(a,b)>0)
        cout<<a<<">"<<b;
    else if (strcmp(a,b)==0)
        cout<<a<<"="<<b;
    else if (strcmp(a,b)<0)
        cout<<a<<"<"<<b;
    cout<<"\n";
}

main(){
    cout<<"***************************************************"<<"\n";
    cout<<"                    字符串的比较                   *"<<"\n";
```

```
            cout<<"*******************************************"<<"\n";
            funArray();
            cout<<"*******************************************"<<"\n";
        }
```

程序运行结果如下：

```
*******************************************
                字符串的比较                    *
*******************************************
请输入字符串（最长 7 位）a=abc
请输入字符串（最长 7 位）b=abc


abc=abc
*******************************************
```

需要指出的是，在比较字符串的大小时，是按照两个字符串顺序逐个字母比较其ASCII码，当一个字符串的一个字母比另一个字符串的相同位置的字母的ASCII码值大时，就认为这个字符串比另外一个大，比较过程结束。否则就一直比较到最后一个字母，如果所有字母都相同，则认为这两个字符串相等。

在运行本例时，读者可以按照上述比较原则，多运行几次，分别输入不同的字符串，看一下程序的运行结果。

5．从字符串中顺序查找字符的函数strchr

strchr函数的作用是，从一个字符串的第一个字符起，顺序查找指定的字符，其函数原型如下：

```
    char *strchr(const char *s, int c);
```

需要注意的是，strchr函数返回的是指定字符在给定字符串中的第一个匹配处，并从这个匹配处开始输出后面的字符串。

　实例6-15　在给定字符串中查找指定字符。

```
    #include <iostream.h>


    void funArray();


    void funArray(){
        char a[8];
        char b;


        cout<<" 请输入源字符串（最长 7 位）:" ;
        cin>>a;
```

```
        cout<<"请输入要查找的字符 :";
        cin>>b;
        cout<<"\n";


        cout<<"第一个匹配处 :"<<strchr(a,b)<<"\n";
        cout<<b<<"在 "<<"'"<<a<<"'"<<"中 第 一 次 出 现 的 位 置 是 第 "<<strlen(a-
strchr(a,b))+1<<"个字母 "<<"\n";
    }


main(){
    cout<<"*******************************************************"<<"\n";
    cout<<"                    字符串的比较                    *"<<"\n";
    cout<<"*******************************************************"<<"\n";
    funArray();
    cout<<"*******************************************************"<<"\n";
}
```

程序运行结果如下：

```
*******************************************************
                    字符串的比较                    *
*******************************************************
请输入源字符串（最长 7 位）:study
请输入要查找的字符 :u


第一个匹配处 :udy
u 在 'study' 中第一次出现的位置是第 3 个字母
*******************************************************
```

6. 从字符串中倒序查找字符函数strrchr

strrchr函数与strchr函数的功能相同，都是从一个给定字符串中查找指定的字符，所不同的是，strchr函数是从字符串的第一个字母开始向后查找，而strrchr函数则相反，是从字符串的最后一个字母开始向前查找，其函数原型与strchr函数的原型也相同。

```
char *strrchr(const char *s, int c);
```

这里不再就此函数举例说明，请读者自行分析。

7. 从字符串中查找指定字符串第一次出现的函数strstr

strstr函数的作用是从一个字符串中查找指定字符串，其函数原型如下：

```
char *strstr(char *str1,char *str2);
```

实例6-16　在字符串中查找指定字符串。

```cpp
#include <iostream.h>


void funArray();


void funArray(){
    char a[50];
    char b[30];


    cout<<"请输入源字符串（最长 49 位）:" ;
    cin>>a;
    cout<<"请输入要查找的字符串（最长 29 位）:";
    cin>>b;
    cout<<"\n";


    cout<<"第一个匹配处:"<<strstr(a,b)<<"\n";
}


main(){
    cout<<"**************************************************"<<"\n";
    cout<<"                    字符串的比较                  *"<<"\n";
    cout<<"**************************************************"<<"\n";
    funArray();
    cout<<"**************************************************"<<"\n";
}
```

程序运行结果如下：

```
**************************************************
                    字符串的比较                 *
**************************************************
请输入源字符串（最长 49 位）:C++魅力无穷！
请输入要查找的字符串（最长 29 位）:魅力


第一个匹配处:魅力无穷！
**************************************************
```

8. 将给定字符串倒转的函数strrev

strrev函数的作用是将给定的字符串倒转，其函数原型如下：

```cpp
char *strrev(char *str);
```

实例6-17 字符串倒转。

```cpp
#include <iostream.h>

void funArray();

void funArray(){
    char a[50];

    cout<<"请输入源字符串（最长49位）:" ;
    cin>>a;
    cout<<"\n";

    cout<<"字符串倒转:"<<strrev(a)<<"\n";
}

main(){
    cout<<"**************************************************"<<"\n";
    cout<<"                     字符串的倒转                    *"<<"\n";
    cout<<"**************************************************"<<"\n";
    funArray();
    cout<<"**************************************************"<<"\n";
}
```

程序运行结果如下：

```
**************************************************
                     字符串的倒转                    *
**************************************************
请输入源字符串（最长49位）:ABC

字符串倒转:CBA
**************************************************
```

9. 把字符串中的小写字母转变成大写字母的函数strupr

strupr函数的作用是将字符串中的小写字母全部转变成大写字母，其函数原型如下：

```cpp
char *strupr(char *str);
```

实例6-18 将字符串中的小写字母转变为大写字母。

```cpp
#include <iostream.h>
```

```
void funArray();

void funArray(){
    char a[50];

    cout<<" 请输入字符串:" ;
    cin>>a;
    cout<<"\n";

    cout<<" 全部都是大写字母:"<<strupr(a)<<"\n";
}


main(){
    cout<<"*********************************************"<<"\n";
    cout<<"*            字符串中的小写字母转变为大写字母            *"<<"\n";
    cout<<"*********************************************"<<"\n";
    funArray();
    cout<<"*********************************************"<<"\n";
}
```

程序运行结果如下:

```
*********************************************
*            字符串中的小写字母转变为大写字母            *
*********************************************
请输入字符串:cBa

全部都是大写字母:CBA

*********************************************
```

实例6-19实现的是将所输入字符串中的大写字母转化为小写的功能,程序中要用到4个函数:isupper、islower、toupper和tolower。其中,isupper用于判断是否为大写字母,islower用于判断是否为小写字母,后面两个则是实现大小写转换的。

实例6-19 将字符串中的大写字母转变为小写字母。

```
// 大写字母小写字母
#include <iostream.h>    // 引入标准 I/O 函数
#include <string.h>      // 引入字符串函数
#include <ctype.h>       // 引入字母测试与转换函数
```

```
int main(){
    char string[] ="Developer Studio";
    cout << "字串转换之前:" << string << endl;     // 显示转换前的字串
    int len = strlen(string);                      // 取得字串长度
    for (int i = 0; i <= len; i++)  {
        if (isupper(string[i]) != 0)               // 若为大写字母
            string[i] = tolower(string[i]);        // 转化成小写字母
    }
    cout << "字串转换之后:" << string << endl;  // 显示转换后的字串
    return 0;
}
```

6.9 上机操作

6.9.1 冒泡法排序

所谓冒泡法排序就是指在给定的x个数中，先将第一个和第二个数相比较，将较大的一个排在第二的位置，再将第二个和第三个数相比较，将较大的一个排在第三的位置。以此类推，直到将第x-1个和第x个数相比较，把较大的放在第x的位置，那么这个数字就是这x个数中最大的一个。之后，再按照同样的方法从第一个开始比较，一直比较到第x-1个，那么第x-1个位置上的数就是这x个数中第二大的数。以此类推，直到把这x个数都按照此顺序排列完毕。

实例6-20　使用冒泡法进行排序。

```
#include <iostream.h>

void funSort();

void funSort(){
    int i,j;
    float t;
    float a[5];

/* 逐个输入数字 */
    for (i=0; i<5; i++)  {
        cout<<"a["<<i<<"]=";
        cin>>a[i];
```

```
        }

/* 利用循环嵌套逐个比较,直到数据排列完毕。*/
    for (i=0;i<5;i++)
        for (j=i+1;j<5;j++)
            if (a[i]<=a[j])   {/* 当第 1 个数小于第 2 个数时,它们交换位置。*/
                t=a[i];  /* 变量 t 用来作为两个数交换位置时的中间交换 " 场地 "*/
                a[i]=a[j];
                a[j]=t;
            }
    cout<<"\n\n";
    cout<<" 排序结果(从大到小)";
    for (i=0;i<5;i++)  /* 顺序输出(即从小到大的顺序)*/
        if (i!=4)
            cout<<a[i]<<"<";
        else if (i==4)
            cout<<a[i]<<"\n";

    cout<<" 排序结果(从小到大)";
    for (i=4;i>=0;i--)  /* 顺序输出(即从小到大的顺序)*/
        if (i!=0)
            cout<<a[i]<<">";
        else if (i==0)
            cout<<a[i]<<"\n";
}


main(){
    cout<<"***********************************************"<<"\n";
    cout<<"*              冒泡法排序               *"<<"\n";
    cout<<"***********************************************"<<"\n";
    funSort();
    cout<<"***********************************************"<<"\n";
}
```

程序运行结果如下:

```
***********************************************

*              冒泡法排序               *
```

```
**********************************************

a[0]=100

a[1]=99

a[2]=101

a[3]=98

a[4]=1109


排序结果（从大到小）98<99<100<101<1109
排序结果（从小到大）1109>101>100>99>98

**********************************************
```

本例中有详细的代码注释，这里不再赘述。

6.9.2 矩阵加法

从数学概念上讲，矩阵是由方程组的系数及常数所构成的方阵。反映在C++程序中，矩阵可以用多维数组来表示，如图6-3所示。

$$X = \begin{pmatrix} 9 & -1 & 8 \\ 11 & 6 & 7 \\ -2 & -5 & 9 \end{pmatrix}$$

图6-3 矩阵

上面这个矩阵由3行3列构成，称为3×3矩阵。

在数学中，行、列数分别对应相同的矩阵可以做加法运算，并且两个矩阵相加后得到的还是一个矩阵。两个矩阵相加的数学规则是：两个矩阵中行列位置相同的元素值相加，得到新矩阵相同位置的元素的值。加法规则如图6-4所示。

$$X\begin{pmatrix} 9 & -1 & 8 \\ 11 & 6 & 7 \\ -2 & -5 & 9 \end{pmatrix} + Y\begin{pmatrix} 15 & 6 & 11 \\ 1 & 9 & -3 \\ 9 & -1 & 9 \end{pmatrix} = Z\begin{pmatrix} 24 & 5 & 19 \\ 12 & 15 & 4 \\ 7 & -6 & 18 \end{pmatrix}$$

图6-4 矩阵加法

实例6-21演示了在C++中对上述矩阵进行加法运算的实现。

实例6-21 矩阵加法运算。

```cpp
#include <iostream.h>


void myFun();


void myFun(){
    int a, b;
```

```
    /* 以二维数组的形式定义矩阵 x、y*/
    int x[3][3]={{9,-1,8},{11,6,7},{-2,-5,9}};
    int y[3][3]={{15,6,11},{1,9,-3},{9,-1,9}};
    /* 定义 x、y 相加后得到的矩阵 */
    int z[3][3];

    /* 此处计算矩阵 z*/
     for (a=0;a<3;a++)  {                    /* 此处的一对大括号可以省略 */
         for (b=0;b<3;b++)
             z[a][b]=x[a][b]+y[a][b];
     }

     for (a=0;a<3;a++)  {                    /* 此处的一对大括号不能省略 */
         for (b=0;b<3;b++)
             cout<<""<<z[a][b];
         cout<<"\n";
     }

     cout<<"\n";
}

main(){
    cout<<"*****************************************"<<"\n";
    cout<<"*                矩阵加法                *"<<"\n";
    cout<<"*****************************************"<<"\n";
    myFun();
    cout<<"*****************************************"<<"\n";
}
```

程序运行结果如下:

```
*****************************************
*                矩阵加法                *
*****************************************
 24  5  19
 12  15  4
 7  -6  18
```

****

在本例中，首先定义了两个二维数组x和y，并分别为它们赋初值，用这两个二维数组分别代表矩阵x和矩阵y；然后又定义了一个二维数组z，用它来代表矩阵z。矩阵的加法操作比较简单，其公式如下：

```
z[a][b]=x[a][b]+y[a][b]
```

即相同位置的矩阵元素的值相加，得到新矩阵的同位置的元素的值。

6.9.3 用顺序法查找指定数值

一个一维数组中有10个元素，根据用户的需求，查找指定数值在该数组中的位置。这个程序比较简单，其主要实现方法是逐个比较数组中的值与用户输入的值是否一致，如果一致则输出该值在数组中的位置，否则给出没有查到该数值的提示信息；最后给出统计信息。

实例6-22 使用顺序法查找指定数值。

```cpp
#include <iostream.h>

void inputNum();
int seekArray(int x,int z);

short int n=0;

void inputNum(){
    int a;
    int y=0,z=0;
    int b;
    int m;

    cout<<" 请输入你要查找的值 :";
    cin>>a;

    for (b=1;b<=10;b++)  {
        m=seekArray(a,z);
        if ((m==-1)&&(n==0))
            cout<<" 该数组中没有你要查找的值 !";
        else if (m==1)  {
            cout<<" 已找到你需要的数值 :"<<"\n";
            cout<<" 位置 :"<<"a["<<b-1<<"]"<<"\n";
```

```
            y++;
        } else  {
          if (b<10)
            cout<<" 继续查找…"<<"\n";
          else
            ;
        }
      z++;
    }
    cout<<" 查找结束，共找到 "<<y<<" 个值。";
    cout<<"\n";
}

int seekArray(int x,int z){
    int i;
    int j=0;
    int myArray[10]={1,2,-5,97,100,110,1,0,11,1};

    i=z;
    if (x==myArray[i])  {
        return 1;
        j++;
    }
     else if (z==9)  {
        n=1;
        return -1;
    } else
        return 0;
}

main(){
    cout<<"***************************************************"<<"\n";
    cout<<"*              在数组中查找指定的数值              *"<<"\n";
    cout<<"***************************************************"<<"\n";
    inputNum();
    cout<<"***************************************************"<<"\n";
}
```

从该程序的运行结果能够清晰地看到程序查找的过程。

当数组中有1个用户要查找的值时，程序运行结果如下：

```
**************************************************
*              在数组中查找指定的数值              *
**************************************************

请输入你要查找的值 :100
继续查找…
继续查找…
继续查找…
继续查找…
已找到你需要的数值 :
位置 : a[4]
继续查找…
继续查找…
继续查找…
继续查找…
查找结束，共找到 1 个值。

**************************************************
```

程序运行结果中已经标明了查到该值时的位置。

当数组中有两个以上用户要查找的值时，程序运行结果如下：

```
**************************************************
*              在数组中查找指定的数值              *
**************************************************

请输入你要查找的值 :1
已找到你需要的数值 :
位置 : a[0]
继续查找…
继续查找…
继续查找…
继续查找…
继续查找…
已找到你需要的数值 :
位置 : a[6]
继续查找…
继续查找…
已找到你需要的数值 :
```

位置：a[9]

查找结束，共找到 3 个值。

＊＊＊＊＊＊＊＊＊＊＊＊＊＊＊＊＊＊＊＊＊＊＊＊＊＊＊＊＊＊＊＊＊＊＊

当程序中没有用户要查找的结果时，则程序运行结果如下：

＊＊＊＊＊＊＊＊＊＊＊＊＊＊＊＊＊＊＊＊＊＊＊＊＊＊＊＊＊＊＊＊＊＊＊

＊　　　　　　　　　在数组中查找指定的数值　　　　　　　＊

＊＊＊＊＊＊＊＊＊＊＊＊＊＊＊＊＊＊＊＊＊＊＊＊＊＊＊＊＊＊＊＊＊＊＊

请输入你要查找的值：1000

继续查找…

继续查找…

继续查找…

继续查找…

继续查找…

继续查找…

继续查找…

继续查找…

继续查找…

查找结束，共找到 0 个值。

＊＊＊＊＊＊＊＊＊＊＊＊＊＊＊＊＊＊＊＊＊＊＊＊＊＊＊＊＊＊＊＊＊＊＊

在程序中，通过一个for循环多次调用函数seekArray，比较用户输入的值与数组中设定的值是否匹配，并且在不同情况下返回不同的值给主调函数。

本题还有另外一种方法：只调用一次seekArray函数。在seekArray函数中，用循环语句来判断用户的输入是否与数组中已经设定的值匹配，如果匹配则返回匹配的值，否则返回特定的值，供主调函数判断并输出信息。此解法请读者自行编程，并亲自上机试验。

6.9.4　用二分查找法查找数组中的值

在6.9.3节中使用的是顺序查找法查找数组中的值，实际上这种方法的效率很低，特别是对大的数组。二分查找法（又叫折半查找法）也是数据查询的一种方法，而这种查找法的效率更高，适用于对大型数组的查询。

使用二分查找法对数组进行查询时，该数组必须是一个排好序的数组（假定按从小到大的顺序排列），其算法如下：

（1）先判断这个数组的中间元素是不是要找的元素，如果是则返回该元素的下标值。

（2）如果不是，再比较这个中间元素的值与要查找的元素的值的大小。

（3）如果大于要查找的元素值，则去该数组的前半部分继续按照本方法查找。

（4）如果小于要查找的元素值，则去该数组的后半部分继续按照本方法查找。

（5）得到查找结果。

实例6-23　二分查找法的应用。

```cpp
#include <iostream.h>

void inputNum();
int binarySearch(int x);

const int NUM=10;
int myArray[NUM]={-1,0,9,10,11,26,70,78,91,101};

void inputNum(){
    int a;
    int z;

    cout<<" 请输人你要查找的数值 :";
    cin>>a;
    z=binarySearch(a);

    if (z==-1)
        cout<<" 没有查到合适的记录 !"<<"\n";
    else  {
        cout<<" 你要查找的记录在数组中的位置是 : "<<"a["<<z<<"]";
        cout<<"\n";
    }
}

int binarySearch(int x){
    int low,high,middle;
    low = 0;
    high = NUM-1;
    while (low<=high)  {
        middle = (low+high)/2;
        if (x==myArray[middle])
            return middle;
        else if (x<myArray[middle])
            high = middle - 1;
        else
```

```
        low = middle + 1;
    }
    return -1;
}


main(){
    cout<<"*****************************************"<<"\n";
    cout<<"*                二分查找法                *"<<"\n";
    cout<<"*****************************************"<<"\n";
    inputNum();
    cout<<"*****************************************"<<"\n";
}
```

程序运行结果如下：

```
*****************************************
*                二分查找法                *
*****************************************

请输入你要查找的数值：10
你要查找的记录在数组中的位置是：a[3]
*****************************************
```

如果数组中没有要查找的值，则程序的运行结果如下：

```
*****************************************
*                二分查找法                *
*****************************************

请输入你要查找的数值：100
没有查到合适的记录！
*****************************************
```

6.9.5　统计学生考试成绩并给出评价

实例6-24　学生成绩统计及评估。

```
#include <iostream.h>
#include <string.h>


int studentMark();
int test(float x);
int eva(float m);
```

```
    int test(float x){
        if (x>100)
            return 1;
        else if (x<0)
            return -1;
        else
            return 0;
    }

    int studentMark(){
        char studentName[10];
        float mark[3];
        float totalMark;
        float avgMark;
        int i;

        cout<<"请输入学生姓名:";
        cin>>studentName;

        cout<<"请输入该生的考试成绩 "<<"\n";
        cout<<"语文:";
        cin>>mark[0];
        cout<<"数学:";
        cin>>mark[1];
        cout<<"外语:";
        cin>>mark[2];

        /* 判断输入的数值是否在范围内 */
        for (i=0;i<=2;i++)  {
            if (test(mark[i])==1)  {
                cout<<"输入错误，成绩不能大于100分！ ";
                cout<<"\n";
                return 1;
            }
            if (test(mark[i])==-1)  {
                cout<<"输入错误，成绩不能为负数！ ";
```

```
            cout<<"\n";
            return 1;
        }
    }

    for (i=0;i<=2;i++)
        totalMark=totalMark+mark[i];
    avgMark=totalMark/3;

    cout<<"学生 "<<studentName<<" 的总成绩:"<<totalMark<<"\n";
    cout<<"学生 "<<studentName<<" 的平均成绩:"<<avgMark<<"\n";

    switch(eva(avgMark))  {
        case 1:cout<<"该生总体评价为:优秀 "<<"\n";break;
        case 2:cout<<"该生总体评价为:及格 "<<"\n";break;
        case 3:cout<<"该生总体评价为:不及格 "<<"\n";break;
        default:cout<<"计算错误 !";break;
    }

    return 0;
}

int eva(float y){
    if (y>=90)
        return 1;
    else if (y>=60||y<90)
        return 2;
    else
        return 3;
}

main(){
    cout<<"******************************************"<<"\n";
    cout<<"*              学生成绩统计程序              *"<<"\n";
    cout<<"******************************************"<<"\n";
    studentMark();
```

```
        cout<<"*******************************************"<<"\n";
}
```

程序运行结果如下：

```
*****************************************
*           学生成绩统计程序            *
*****************************************
请输入学生姓名：张晓华
请输入该生的考试成绩
语文:99
数学:100
外语:98
学生 张晓华 的总成绩:297
学生 张晓华 的平均成绩:99
该生总体评价为：优秀
*****************************************
```

本例用数组studentName来存放学生的名字，用数组mark来存放学生的三个学科的成绩，用函数test来验证所输入的学生成绩是否在0~100范围内，并根据输入错误的不同给出不同的错误提示。接下来计算学生的三科总成绩和平均成绩，再根据平均成绩来调用函数eva给出对该学生的评价。

6.9.6 区分字符数组中的大小写字母

实例6-25 提取字符串数组中的大小写字母。

```
#include <iostream.h>
#include <string.h>


void myFun();


void myFun(){
    char a[10]={'a','A','b','B','c','C','d','D','e','E'};
    char b[5];
    char c[5];
    int x,m,n;


    m=0;
    n=0;
    for (x=0;x<10;x++)  {
```

```
            if (isupper(a[x]))  {
                b[m]=a[x];
                m++;
            }
            else if (islower(a[x]))  {
                c[n]=a[x];
                n++;
            }
            else
                continue;
        }

    cout<<" 原始数组：";
    for (x=0;x<10;x++)
        cout<<a[x];
    cout<<"\n";

    cout<<" 大写数组 ：";
    for (x=0;x<5;x++)
        cout<<b[x];
    cout<<"\n";

    cout<<" 小写数组 ：";
    for (x=0;x<5;x++)
        cout<<c[x];
    cout<<"\n";
}


main(){
    cout<<"*******************************"<<"\n";
    cout<<"*     区分数组中的大小写字母     *"<<"\n";
    cout<<"*******************************"<<"\n";
    myFun();
    cout<<"*******************************"<<"\n";
}
```

程序运行结果如下：

```
****************************
*     区分数组中的大小写字母     *
****************************
原始数组：aAbBcCdDeE

大写数组：ABCDE

小写数组：abcde

****************************
```

在本例中给字符数组a赋予的初值中有大写字母也有小写字母，目的是要把这些大写字母和小写字母区分开来，并分别写入另外两个字符数组中。需要注意的是，在本例中使用了两个C++的库函数isupper和islower，这两个函数是C++提供的用于判断字母是否是大写字母和小写字母的函数，在6.8节中已应用过这两个函数。

6.10 小结

本章讲述了C++中两个重要的数据类型：数组和字符串。数组是把具有相同数据类型的变量按照一定的次序组织起来的数据类型。按照数组元素的数据类型，可以将数组分成字符数组、整型数组、浮点型数组等。按照数组的维数，可以将数组分成一维数组、二维数组和多维数组。在C++中，字符串是以字符数组的形式存在的。本章配合实例讲解了几个实用的字符串处理函数，请读者多多练习。

6.11 习题

一、填空题

1．数组是C++的一种数据类型，它的使用遵循C++中变量的使用规则，即：_____。

2．C++访问数组元素是通过数组的_____值来进行的，而C++本身并不对数组的_____进行检查。

3．数组a[i]共有_____个数组元素，其中第一个元素是a[__]。

4．数组a[]的元素分别为{-100,101,0,99,-1}，那么a[0]=_____，a[3]=_____。

5．有一个二维数组a[][]，其元素分别为{{10,11,-1}, {99,101,-1}, {0,1,2,}, {123,-100,0}, {99,-2,97}}，这个数组是一个_____行_____列的数组，其中a[1][2]=_____。

6．如果数组a[]="C++"，那么strlen(a)=_____。

二、程序阅读题

1．下面是一个使用二分法查找数组元素的程序，请根据所学知识将这个程序补充完整。

```
#include <iostream.h>
```

```cpp
void inputNum();
int binarySearch(int x);


const int NUM=10;
int myArray[NUM]={-200,-101,-7,0,1,2,90,97,109,234};


void inputNum(){
    int a;
    int z;

    cout<<"请输入你要查找的数值:";
    cin>>a;
    z=binarySearch(a);

    if (_____)
        cout<<"没有查到合适的记录!"<<"\n";
    else {
        cout<<"你要查找的记录在数组中的位置是:"<<"a["<<_____<<"]";
        cout<<"\n";
    }
}


int binarySearch(int x){
    int low,high,middle;
    low = 0;
    high = NUM-1;
    while (_____) {
        middle = _____;
        if (x==myArray[middle])
            _____;
        else if (_____)
            high = middle - 1;
        else
            low = middle + 1;
    }
    return -1;
```

```
}

main(){
    cout<<"*******************************************"<<"\n";
    cout<<"*            二分查找法查找数组元素          *"<<"\n";
    cout<<"*******************************************"<<"\n";
    inputNum();
    cout<<"*******************************************"<<"\n";
}
```

2．求杨辉三角。

杨辉三角是一个由数字排列成的形状像三角形的表，其形状如下：

```
        1
      1   1
    1   2   1
  1   3   3   1
1   4   6   4   1
1  5  10  10  5  1
```

从上面的数字可以看到：杨辉三角的每条斜边都是1，其余的数则是等于它肩上的两个数之和。

下面这个程序就是用来求杨辉三角的，请读者根据程序运行结果分析数组的应用和格式控制。程序源代码如下，请将其补充完整。

```
#include <iostream.h>

void yHui();

const int N=10;

void yHui(){
    int yTriangle[N],row,col;

    yTriangle[0]=1;
    cout<<yTriangle[0]<<"\n";
    for (row=1;row<=_____;row++)  {
        yTriangle[row]=1;

        for(col=_____;col>0;col--)
```

```
            yTriangle[col]=yTriangle[col]+yTriangle[col-1];


        for(col=0;col<=row;col++)

            cout<<yTriangle[col]<<"";


        cout<<"\n";
    }
}


main(){
    cout<<"*****************************************"<<"\n";
    cout<<"*                    杨辉三角                    *"<<"\n";
    cout<<"*****************************************"<<"\n";
    yHui();
    cout<<"*****************************************"<<"\n";
}
```

程序运行结果如下：

```
*****************************************
*                    杨辉三角                    *
*****************************************
                      1
                    1   1
                  1   2   1
                1   3   3   1
              1   4   6   4   1
            1   5   10   10   5   1
          1   6   15   20   15   6   1
        1   7   21   35   35   21   7   1
      1   8   28   56   70   56   28   8   1
    1   9   36   84   126   126   84   36   9   1
  1 10   45   120   210   252   210   120   45   10   1
*****************************************
```

三、问答题

1. 数组和普通变量有何异同？

2. 数组可以分为哪几种？

3. 什么是C++的字符串？

第7章 指 针

本章学习目标

➤ 指针的概念　　　　　➤ 指针变量

➤ 指针和内存地址的关系　➤ 数组指针和指向数组元素的指针变量

前面已讲过了C++的多种数据类型，例如：整型、浮点型、字符型、数组等。本章将讲述C++另一种非常重要的数据类型——指针。指针在C++中是一种极其重要的数据类型，正是指针使得C++的程序更加灵活、简洁、紧凑和高效。特别对于系统软件来说，指针能够解决一些使用通常方法所不能解决的问题。

但是，因为指针极具灵活性，这导致它的知识结构、使用方法也非常复杂，指针在给C++程序带来巨大便利的同时，也为C++程序带来了更多的复杂性。

7.1 指针的概念

指针即变量的地址，它是变量在内存中的存储地址，通过它可以找到这个变量。为了说明指针概念的含义，需要先熟悉下面三个知识点：内存地址、变量地址和变量值的存取。

1. 内存地址

一个住宅小区由很多座居民楼组成，这些居民楼又由许多单元组成，每个单元都有很多的房间，房间内住着居民。为了方便管理，小区中的每一座楼、每个单元、每个房间都被编号了，而且这些编号在这个小区中是唯一的。比如：小明家在3号楼2单元301房间，小华家在9号楼6门302房间等。

计算机的内存也是这样管理的。计算机内存的数据存储部分由很多存储单元组成，每个存储单元能存储1字节的数据量，计算机系统为每个存储单元分配了在内存中唯一的编号，这个编号就是存储单元在内存中的地址。存储单元中的"住户"正是变量、常量或常数等。图7-1表示了内存地址的编号。

5007	5008	5009	5010	5011

图7-1　内存地址编号

图7-1中5007、5008、5009、5010和5011代表内存中的5个存储单元，而这5个数字就是5个内存单元的地址，它们在内存中是唯一的。但是，它们仅仅是地址，就像我们家门上贴着的门牌一样，它们并不是存储在存储单元中的数据，图7-2表示了内存地址和数据的关系。

| 1 | 2 | A | a | B |

图7-2　内存地址与存储数据

从图7-2可以看到，内存地址标识出数据在内存中存放的位置，比如存储单元5007中存放的数据是1，内存单元5010中存放的数据是a等。因为内存地址是唯一的，所以可以通过内存地址快速、方便地找到变量。

2. 变量地址

上面讲到了内存地址，变量地址就是变量在内存中存放的首地址。因为不同数据类型的变量所描述的数据范围是不同的，例如：int型占用4个字节，而short int型占用两个字节。这意味着在内存的一个单元中不能存放一个int型的数据，怎么办呢？答案很简单，可以将数据存放在不同的单元中，存取这个变量时只需要知道数据在哪些单元中存放的和存放的顺序就可以了。

如何知道数据到底存放在哪些单元呢？又是按什么顺序排列呢？内存中有一段称为符号表，在符号表中存放了变量的标识符和变量在内存中的地址，只要对照符号表就能够找到变量，并能够正确地对变量进行读写。

提示

变量地址并不是这个变量在内存中存放的每个内存地址，而仅仅是变量在内存中存放的首地址。一个变量在内存中到底存放在哪些内存地址中，则是通过符号表来表示的。

请看下面的程序段：

```
#include <iostream.h>

main(){
  …
  int a;
  cin>>a;
  cout<<a<<"\n";
  …
}
```

C++编译系统在遇到上面的变量a的定义时，就为a在内存中开辟4个内存存储空间，并把这4个存储空间以及第一个空间的地址登记在符号表中，因此就可以很方便地利用cin语句和cout语句对变量a进行读写操作了。

3. 变量值的存取

知道了什么是内存地址，什么是变量地址，那么变量在内存中是如何存取的呢？换句话讲，变量在内存中如何被访问呢？

访问变量可以有两种方法。

• 直接访问法　通过变量地址直接对变量进行读写操作。在上面的程序段中定义了一个int型变量a，变量a的首地址已经被登记到符号表中了。因此，通过访问符号表就能够知道这个变量在内存中的存储地址，并按照这个地址找到变量，对它进行读或写操作。

• 间接访问法 不同于直接访问法，C++规定了一种特殊的变量，这种变量用来存放其他变量的地址，通过这种变量间接地找到需要寻找的变量，这种变量被称为指针变量。

在变量名的前面加上一个*号就使其变成了指针变量，指针变量和其他类型的变量一样，都必须有数据类型，这个数据类型指定了该指针变量要指向的数据类型。例如：把一个指针变量定义为int型，那么通过该指针变量存取的将是一个整数。同样，把一个指针变量定义为char型，那么通过该指针变量存取的将是一个字符型。指针所指向数据的类型通常也称为指针的类型。

假设有一个字符型的变量a，它的值为'A'.编译器在编译到这条定义语句时，将为它分配内存空间，此空间的内存地址为p，此时就可以说p就是指向变量a的指针，通过指针p可以间接地访问到变量a，并对变量a进行读写操作。图7-3为指针与其指向的数据之间的关系。

指针 p   变量 a

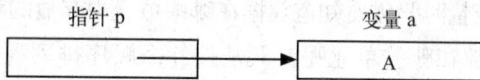

图7-3 指针与指向数据之间的关系

7.2 指针变量

在上一节中讲过，C++有一类特殊变量，可以用它来存放其他变量的地址，这种变量称为指针变量。指针变量的定义格式如下：

```
<数据类型> *<指针变量名>;
```

可以看到，指针变量的定义方式与普通变量是一致的，只是要在指针变量名前加上一个星号，表示后面跟的是一个指针变量。例如：

```
int *p;

char *a;

double *pa;
```

这3条语句分别定义了整型指针变量、字符型指针变量和浮点型指针变量。

C++中有一种操作符&，叫做取地址操作符，用来取得变量的首地址。例如&a，表示取得变量a在内存中存储单元的首地址。由此可见：指针变量用来存储变量的地址，而&操作符用来取得变量的首地址，可以把&操作符取得的地址赋给指针变量。例如：

```
int *p=&a;

char *a=&b;

double *pa=&x;
```

这三条语句的解释如下：

（1）第一条语句首先定义了一个整型指针变量p，然后用取地址操作符&取得变量a的地址，并将这个地址赋给指针变量p。

（2）第二条语句首先定义了一个字符型指针变量a，然后用取地址操作符&取得变量b的地址，并将这个地址赋给指针变量a。

（3）第三条语句首先定义了一个浮点型指针变量pa，然后用取地址操作符&取得变量x的地址，并将这个地址赋给指针变量pa。

　　这三条语句的定义方式是一样的，只是指针变量的数据类型不同罢了。在定义指针变量时，除了可以使用普通的数据类型外，还可以使用void关键字，用该关键字定义的指针变量表示此指针变量无类型，或者说它仅仅指向内存的一个存储单元，但并没有指向一个具体的数据类型。

　　下面这个例子演示了指针变量的定义和使用。

实例7-1　指针变量的定义和使用。

```
#include <iostream.h>

void pointFun();

void pointFun(){
    int a=10, *pa;
    float b=1.5, *pb;
    char c='x', *pc;


    pa=&a;
    pb=&b;
    pc=&c;


    cout<<"a="<<a<<";    pa="<<*pa<<"\n";
    cout<<"b="<<b<<";    pb="<<*pb<<"\n";
    cout<<"c="<<c<<";    pc="<<*pc<<"\n";
}


main(){
    cout<<"*********************************"<<"\n";
    cout<<"*        指针变量的定义和使用        *"<<"\n";
    cout<<"*********************************"<<"\n";
    pointFun();
    cout<<"*********************************"<<"\n";
}
```

程序运行结果如下：

```
*********************************
*        指针变量的定义和使用        *
*********************************
a=10;    pa=10
```

```
b=1.5;    pb=1.5

c=x;    pc=x

**********************************
```

在这个程序中，定义了3个指针变量pa、pb和pc，它们被定义后并没有说明要指向哪个地址，这时候的指针变量并不指向一个特定的地址，所以处于空指针状态。接下来，用取地址操作符&分别取得3个变量a、b和c的地址，并将它们的地址分别赋给3个指针变量，这时的指针变量不再处于空指针状态了，它们分别有了特定的指向目标。

7.3 指针变量赋值

在7.2节里讲了指针变量的概念，一个指针被定义后应该被赋予相应的值，也就是要让它指向一个地址，否则指针就是空指针，而空指针对于系统是有害的。因此，需要为已定义的指针变量赋值。

为指针变量赋值的格式如下：

指针变量 =& 要指向的变量；

在实例7-1中，首先用取地址操作符取得变量a、b和c的地址，然后通过赋值语句将取得的地址赋给指针变量，程序中的赋值语句如下：

```
pa=&a;

pb=&b;

pc=&c;
```

为指针变量赋值时应该注意以下几点：

（1）被定义的指针变量应该让它指向一个地址，应避免对系统有害的空指针存在。

（2）指针变量只能存放指针，即内存地址。

（3）一个指针变量只能存放相同类型变量的地址。

下面是一个为指针变量赋值的例子。

实例7-2 为指针变量赋值。

```
#include <iostream.h>


void myFun();


void myFun(){
    int a;
    int *pa,*pb,*pc;
    a=10;
    pa=&a;
    pb=&a;
    pc=pb;
```

```
    cout<<"pa="<<*pa<<"\n";

    cout<<"pb="<<*pb<<"\n";

    cout<<"pc="<<*pc<<"\n";

}


main(){

    cout<<"*****************************"<<"\n";

    cout<<"*          指针变量的赋值           *"<<"\n";

    cout<<"*****************************"<<"\n";

    myFun();

    cout<<"*****************************"<<"\n";

}
```

程序运行结果如下：

```
*****************************
*          指针变量的赋值           *
*****************************
pa=10
pb=10
pc=10
*****************************
```

在这个例子中，首先定义了一个整型变量a以及3个指针变量：pa、pb和pc；然后为变量a赋值整数值10，并将指针变量pa、pb都指向变量a的地址。通过计算结果可以看到，指针变量pa和pb所指向的值是一样的。对于指针变量pc，将它的地址指向指针变量pb，因为指针变量pb的值就是变量a的地址，所以指针变量pc的值也应该是变量a的地址。程序运行结果也验证了这一点。

图7-4表示了本例中3个指针变量所指向地址的情况。

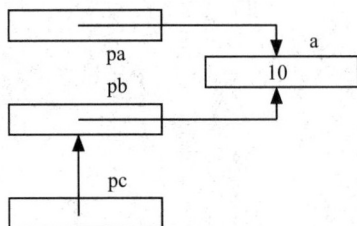

图7-4 指针变量与其指向的地址

7.4 指针运算

指针也可以像普通变量一样进行各种运算，例如赋值运算、比较运算、加减法运算、自增运算、自减运算等。除此之外，指针还有一些自身特有的运算，比如在前面讲到的取地址、间接访问等运算。

1．赋值运算

指针的赋值运算在7.3节中已经讲过，再举例如下：

```
float a=1.0;
float *pa
pa=&a;
```

2．取地址运算

取地址运算符在前面已讲到，就是用&符号取得变量的地址。例如：

```
char c='A';
char *pc;
pc=&c;
```

3．加减法运算

指针变量中存放的是变量的地址，那么地址加n自然就是由现在的地址向前第n个数据的地址了，地址减n自然就是由现在的地址向后第n个数据的地址。

例如：

```
char a[5]="ABCDE";
char *pa,*pb;
*pa=a;
pb=pa+1;
```

指针pa指向的地址的值应该为A，而pb的值是pa的值加1，那么pb的值应该是pa的值向后加1，也就是pa指向的地址向后移动1位，即pb指向的地址的值应该为B。

下面这个例子说明了指针变量的加减法。

实例7-3 指针变量的加减法。

```
#include <iostream.h>

void myFun();

void myFun(){
    char myArray[10]="Chinese";
    char *pa=myArray,*pb,*pc;

    pb=pa+3;
    pc=pb-2;
```

```
        cout<<" 完整数组如下 :"<<"\n";

        cout<<myArray<<"\n";

        cout<<"pa="<<*pa<<"\n";

        cout<<"pb=pa+3="<<*pb<<"\n";

        cout<<"pc=pb-2="<<*pc<<"\n";

}


main(){

        cout<<"*****************************"<<"\n";

        cout<<"*          指针变量的加减运算          *"<<"\n";

        cout<<"*****************************"<<"\n";

        myFun();

        cout<<"*****************************"<<"\n";

}
```

程序运行结果如下：

```
*****************************

*          指针变量的加减运算          *

*****************************

完整数组如下 :

Chinese

pa=C

pb=pa+3=n

pc=pb-2=h

*****************************
```

在这个例子中，首先定义了一个字符数组myArray，并为其赋值为Chinese，然后又定义了3个指针变量pa、pb和pc，并为pa赋值为myArray（关于数组的指针我们会在后面几节中讲到），pb=pa+3就是使指针变量pb指向的地址为指针变量pa指向的地址向后移动3位，即为字母n。之后pc=pb-2就是使指针变量pc指向的地址为指针变量pb指向的地址向前移动2位，即为字母h。

4．比较运算

指针变量同普通变量一样，也可以做比较运算。因为指针变量存储的是地址，所以很容易理解地址大的指针变量自然比地址小的指针变量要大。

下面这个简单的例子说明了指针变量的比较运算。

实例7-4　指针变量的比较运算。

```
#include <iostream.h>


void myFun();
```

```
void myFun(){
    char myArray[10]="Hello";
    char *pa=myArray,*pb

    pb=pa+3;

    cout<<"pa="<<*pa<<"\n";
    cout<<"pb=pa+3="<<*pb<<"\n";

    if (pa>pb)
        cout<<*pa;
    else
        cout<<*pb;

    cout<<"\n";
}


main(){
    cout<<"********************************"<<"\n";
    cout<<"*        指针变量的比较运算         *"<<"\n";
    cout<<"********************************"<<"\n";
    myFun();
    cout<<"********************************"<<"\n";
}
```

程序运行结果如下：

```
********************************
*        指针变量的比较运算         *
********************************
pa=H
pb=pa+3=l
l
********************************
```

在本例中，首先定义了一个数组myArray，同时为这个数组赋了初值；之后定义了两个指针变量pa和pb，其中，指针pa指向数组myArray的首地址，也就是myArray[0]的位置，指针pb指向的位置是在指针pa的基础上加3个单位，也就是myArray[3]的位置。从程序运行结果来看，pb的值是大于pa的值。

5．自增1（++）和自减1（--）运算

指针变量同样可以使用自增1运算符++和自减1运算符--。例如：定义一个指针变量p，然后通过p++，使得指针p指向原来数据的后一位数据。使用p--使得指针p指向原来数据的前一位数据。

实例7-5 指针变量的自增自减运算。

```cpp
#include <iostream.h>


void myFun();

void myFun(){

    char myArray[20]="HelloWorld";

    char *pa=myArray;


    cout<<" 原始值:"<<*pa<<"\n";

    cout<<" 自加1值:"<<*pa++<<"\n";

    cout<<" 奇怪！怎么没加？ "<<"\n";

    cout<<" 这样才加 *++pa: "<<*++pa<<"\n";

    cout<<" 也是这样才减的 *--pa: "<<*--pa<<"\n";

}


main(){

    cout<<"*****************************"<<"\n";

    cout<<"*        指针变量的比较运算        *"<<"\n";

    cout<<"*****************************"<<"\n";

    myFun();

    cout<<"*****************************"<<"\n";

}
```

程序运行结果如下：

```
*****************************
*        指针变量的比较运算        *
*****************************
原始值:H
自加1值:H
奇怪！怎么没加？
这样才加 *++pa: l
```

也是这样才减的 *--pa: e

****

在本例中，首先定义了一个字符数组myArray，并为它赋值字符串HelloWorld；然后又定义了一个指针变量pa，并且使指针pa指向数组myArray，这时pa指向的值是首字符H。*pa++操作可以使指针向后移动一位，但是通过语句cout<<"自加1值:"<<*pa++<<"\n";得到的输出值仍然是字符H，为什么？

答案很简单，自增运算符先使用变量的原值，然后才使变量值加1，所以用cout语句输出变量值的时候，事实上输出的还是指针p指向的原值，然后才使指针变量值加1，也就是将指针向后移动一位。而在使用cout<<"这样才加*++pa: "<<*++pa<<"\n";语句时，指针已经指到了第三个字符l了。这是因为表达式*++p会先使指针变量值加1，然后得到指针变量的当前值。使用自减运算--也是同样的道理。

7.5　指针数组与指向数组的指针

数组是由相同数据类型的一组数据元素组成的，那么数组元素是否可以是指针类型呢？答案是肯定的。

7.5.1　指针数组

如果数组中的元素都是由指针变量组成的，那么这个数组的类型就是指针类型，这种数组称为指针数组。例如：

```
int *p[10];

char *p[10];

float *p[10];
```

这些都是指针数组，第一个语句定义了一个整型指针数组，这个数组中所有的数组元素都是指向整型量的指针。第二个语句定义了一个字符型指针数组，这个数组中所有的数组元素都是指向字符型量的指针。第三个语句定义了一个浮点型指针数组，这个数组中所有的数组元素都是指向浮点型量的数组。在这些指针数组中，p[0]、p[1]、p[2]…p[9]都是指针变量，p为指针数组元素p[0]的地址，p+i为p[i]的地址，*p的值为p[0]，*(p+i)的值为p[i]。

前面曾经讲过，数组是通过其下标变量来引用的，例如，数组定义：

```
int myArray[10];
```

通过下标变量来引用，即为：myArray[0]、myArray[1]、myArray[2]、myArray[3]…myArray[9]。事实上，也可以通过指向该数组的指针来引用数组元素，因为指针方式更快、更灵活、占用的系统资源更少，所以使用指针来引用数组元素是一种高效、灵活的方法。

7.5.2　指向数组的指针

与指向普通变量的指针一样，指向数组的指针的定义方式如下：

```
char myArray[10];
```

```
char *p=myArray;
```

或者：

```
char myArray[10];

char *p;

p=myArray;
```

这两种定义方式是一样的，实际上在本章前面几节的例子中已经用到过这种定义了。需要注意的是，可以用数组名来为指针变量赋值，但是数组名仅代表数组的首地址，并不代表整个数组。因此当使用这种方法来为指针变量赋值时，只是把数组的首地址赋给了指针变量。但是，因为数组在内存中是连续存放的，所以可以通过移动指针方便地访问整个数组的所有元素。

实例7-6 指向数组的指针。

```
#include <iostream.h>

void pointFun();

void pointFun(){
    int myArray[5];

    int *p;

    int i;

    p=myArray;
    /* 利用指针为数组赋值 */
    for (i=0;i<5;i++)  {
        cout<<" 请输入 a["<<i<<"] 的值 : ";

        cin>>*p++;
    }
    /* 利用指针输出数组元素 */
    cout<<"a[5]={";

    for (i=0;i<5;i++)  {
        cout<<*(p+i);

        if (i<4)
            cout<<",";

    }
    cout<<"}";

    cout<<"\n";
```

```
    }

    main(){
        cout<<"*********************************"<<"\n";
        cout<<"*          指向数组的指针         *"<<"\n";
        cout<<"*********************************"<<"\n";
        pointFun();
        cout<<"*********************************"<<"\n";
    }
```

程序运行结果如下：

```
*********************************
*          指向数组的指针         *
*********************************
请输入 a[0] 的值：-9
请输入 a[1] 的值：0
请输入 a[2] 的值：97
请输入 a[3] 的值：99
请输入 a[4] 的值：101
a[5]={-9,0,97,99,101}
*********************************
```

本例通过指针变量为数组赋值，又通过指针变量输出整个数组的所有元素，请读者自行分析程序。

7.6 字符串指针与指向字符串的指针

在前面讲字符串时曾经讲过，可以用数组的形式来处理字符串，用数组的下标变量可以表示字符串中的单个字符，例如：

```
myArray[]="Hello World";
```

用数组myArray来存放字符串"Hello World"，并可以用带下标的数组元素来表示该字符串的每个字符，例如：myArray[0]代表字符H，myArray[1]代表字符e。需要注意的是，当字符串"Hello World"被存储到数组myArray中时，最后自动加了一个字符\0，表示字符串结束。

除了用数组来处理字符串外，还有另外一种方法：指针。通过指针引用字符串更方便、更快捷。所谓字符串的指针就是指字符串在内存中的起始地址。例如：定义一个字符指针pc，让pc指向一个字符串，实际上就是指向这个字符串的首地址；然后通过对指针的移动来引用这个字符串。

```
char *pc;
```

```
pc="Hello";
```

这个指针指向该字符串的情况可以用图7-5来表示。

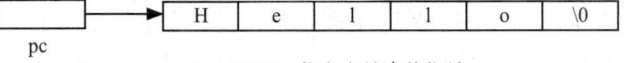

图7-5　指向字符串的指针

使用指针来引用字符串的好处是，可以根据实际情况方便地改变字符串的长度，这一点用数组就做不到了，因为数组在定义时必须指定数组的长度，然后编译系统根据数组的长度定义来为数组分配存储空间，这个空间是静态的，一旦分配完成就不能再更改。因此，使用指针引用字符串更加灵活。

实例7-7　指向字符串的指针。

```
#include <iostream.h>


void myFun();


void myFun(){
    char *pc;


    pc="Hello";
    while (*pc!='\0')  {
            cout<<*pc;
            *pc++;
    }


    cout<<"\n";
}


main(){
    cout<<"***************************"<<"\n";
    cout<<"*      指向字符串的指针    *"<<"\n";
    cout<<"***************************"<<"\n";
    myFun();
    cout<<"***************************"<<"\n";
}
```

程序运行结果如下：

```
***************************

*      指向字符串的指针    *

***************************
```

```
Hello

****************************
```

这是一个简单的指向字符串的指针的示例程序。在程序中先定义了一个字符指针pc，并将它指向预设的字符串"Hello"，然后通过一个while循环来移动指针，并逐个输出字符串的每个字符，直到遇到字符\0为止。

实例7-8　按照名字的拼音顺序排列著作的编委。

```
#include <iostream.h>
#include <string.h>

void init();
void sort(char *name[],int n);
void OutScreen(char *name[],int n);

/* 初始化数组并调用其他函数 */
void init(){
    int x=6;

    void sort(char *name[],int n);
    void outScreen(char *name[],int n);
    char *name[]=
        {"LiGang","LiuXiaoMing","WangCunGuang",
         "LiShunLi","ZhaoYong","QianDuoDuo"
        };

    sort(name,x);
    OutScreen(name,x);
}

/* 排序过程 */
void sort(char *name[],int n){
    char *p;
    int i,j,k;

    for(i=0;i<n-1;i++)  {
      k=i;
      for(j=i+1;j<n;j++)
```

```
        if(strcmp(name[k],name[j])>0)
            k=j;
        if(k!=i)  {
            p=name[i];
            name[i]=name[k];
            name[k]=p;
        }
    }
}


/* 输出到屏幕 */
void OutScreen(char *name[],int n){
    int x;

    cout<<" 编委名单如下 : "<<"\n";
    for (x=0;x<n;x++)  {
        cout<<x+1<<":";
        cout<<name[x]<<"\n";
    }
}


main(){
    cout<<"*****************************************"<<"\n";
    cout<<"*   编委名单 (以名字的汉语拼音为序)   *"<<"\n";
    cout<<"*****************************************"<<"\n";
    init();
    cout<<"*****************************************"<<"\n";
}
```

程序运行结果如下 :

```
*************************************

*   编委名单 (以名字的汉语拼音为序)   *

*************************************

编委名单如下 :

1:LiGang

2:LiShunLi

3:LiuXiaoMing
```

```
    4:QianDuoDuo

    5:WangCunGuang

    6:ZhaoYong

    *****************************************
```

对于字符串的比较及排序，通常会这样做：先比较前两个字符串的大小，将大的通过字符串复制函数放到第一位，再比较第二个字符串和第三个字符串，再把较大的放到第二位。依次类推，最终得到一个按照大小顺序排列的字符串组。这种方法是正确的，只是以前是用数组来完成这种工作，这样无法避免数据被反复拷贝，既增加了系统资源的开销，又降低了程序的执行效率。

在上面这个例子中依然使用的是相同的解题思路，但是实现方法却不同。本例中是把字符串存放在一个数组中，把这些字符数组的首地址放在一个指针数组中，当需要交换两个字符串时，只须交换指针数组相应两个元素的内容（即地址）即可，而不必交换真实的字符串本身。

与数组方式相比，使用指针方式进行比较与排序遵循了相同的解题思路，但是实际交换的并不是真实的数据，而仅仅是一个地址，所以节约了系统时间的开销，对于比较长的字符串更是如此。

7.7　指针函数

指针函数就是返回值为指针类型的函数。就像函数可以返回整型值、浮点型值、字符型值一样，作为C++的一种数据类型，函数同样可以返回指针类型的值，这种函数就是指针函数。指针函数的定义格式如下：

函数类型 ＊函数名()；

实例7-9　利用指针函数比较大小。

```cpp
#include <iostream.h>


void inputNum();

int *myFun(int *x,int *y);


void inputNum(){

    int a,b;

    int *p;


    cout<<"a=";

    cin>>a;

    cout<<"b=";

    cin>>b;
```

```
    p=myFun(&a,&b);
    cout<<"Max="<<*p<<"\n";
}


int *myFun(int *x,int *y){
    if(*x>*y)
        return  x;
    else
        return  y;
}


main(){
    cout<<"*************************************"<<"\n";
    cout<<"*       利用指针函数比较数值大小        *"<<"\n";
    cout<<"*************************************"<<"\n";
    inputNum();
    cout<<"*************************************"<<"\n";
}
```

程序运行结果如下：

```
*************************************
*       利用指针函数比较数值大小        *
*************************************
a=1
b=2
Max=2
*************************************
```

本例比较简单，仅用一个指针函数myFun来做数值的比较，只是它的形参列表中是两个指针类型的参数，因此需要调用它的函数传递给它两个指针类型的参数，然后它再根据传递过来的指针找到真实的数据做比较，将数据较大的那个数值的指针返回给主调函数，再由主调函数完成输出。

实例7-10　抽检轴承检验是否合格。

```
#include <iostream.h>


void init();
float *seek(float (*pnt_row)[3]);
```

```cpp
/* 初始化数组, 并调用检验函数。*/
void init(){
    float numsMark[3][3]={{0.91,0.81,0.70},{1.10,0.89,0.85},{0.5,0.51,0.52}};
    int i,j;
    float *p_r;

    for(i=0;i<3;i++)  {
        p_r=seek(numsMark+i);
        if(p_r==*(numsMark+i))  {
            cout<<" 轴承系数 : "<<i+1<<"\n";
            cout<<"{";
            for(j=0;j<3;j++)  {
                cout<<*(p_r+j);
                if (j<2)
                    cout<<",";
            }
            cout<<"}";
            cout<<"\n";
        }
    }
}


float *seek(float  (*pnt_row)[3]){
    int i=0;
    float *pnt_col;

    pnt_col=*(pnt_row+1);
      for(; i<3; i++)
        if(*(*pnt_row+i)>0.90){
            pnt_col=*pnt_row;
            break;
        }
    return(pnt_col);
}
```

```
main(){
    cout<<"***********************************"<<"\n";
    cout<<"*          轴承指标抽检验证模块          *"<<"\n";
    cout<<"***********************************"<<"\n";
    init();
    cout<<"***********************************"<<"\n";
}
```

程序运行结果如下：

```
***********************************
*          轴承指标抽检验证模块          *
***********************************
轴承系数:1
{0.91,0.81,0.7}
轴承系数:2
{1.1,0.89,0.85}
***********************************
```

这个程序用来验证某种规格的轴承的误差，它的作用是验证每组3个待检测误差中大于0.90的组，如果存在这样的组则将其输出。在本程序中，初始化时给出了固定的三组参数，并将其存储在一个float型数组中。在实际工作中，这一步可能不同，比如会要求用户输入相关数据，这就需要开发一个人机交互的界面，请读者自己实践。

7.8　main函数的参数

前面已经讲过，一个C++程序是由main函数开始执行的，而且一个程序中只能有一个main函数，main函数可以调用任何自定义函数，但是自定义函数却不能调用main函数。

操作系统中提供的实用工具，比如：ping命令，它们是可以带参数运行的，可以用ping ipaddress -t的格式，其中-t就是一个参数。与此类似，我们开发的程序也可以有这种格式，用来完成不同的任务。

在前面所演示的例子中，main函数都是没有形参的。事实上，main函数和其他函数一样也可以带有形式参数的，所不同的是自定义函数的形式参数是用来接收其他函数调用它时传递过来的实参，而main函数的形参主要用来接收执行该C++程序时传递的参数。

需要注意的是，main函数的参数不是随便定义的，它有自己的格式，带有参数的main函数的一般形式如下：

```
int main( void ){}    // 无须从命令行获取参数
int main( int argc, char *argv[] ){}
```

其中参数argc用来说明参数个数，这个参数统计包含程序名本身。参数argv是一个指针数组，用来表明每个参数（随便提示一下：main函数的返回值类型必须是int，这样返回值才能被传递给程序的激活者，比如操作系统）。下面举一个简单的例子，示例代码如下：

```
#include <iostream.h>

void myFun(char *b,int x,int y);

void myFun(char *b,int x,int y){
    int max,min;

    char c;
    c=*b;

    cout<<"*b="<<*b<<"\n";
    cout<<"c="<<c<<"\n";

    if (x>=y)  {
        max=x;
        min=y;
    } else  {
        max=y;
        min=x;
    }

    if (c=='0')
        cout<<" 最大值: "<<max<<"\n";
    else if (c=='1')
        cout<<" 最小值: "<<min<<"\n";
    else
        cout<<" 参数不正确!"<<"\n";
}

main(int argc,char *argv[]){
    int m,n;
    int i;
    cout<<"****************************"<<"\n";
    cout<<"*    带有参数的 main 函数      *"<<"\n";
    cout<<"****************************"<<"\n";
```

```
        cout<<"m=";
        cin>>m;
        cout<<"n=";
        cin>>n;

        cout<<"argc="<<argc<<"\n";
        for (i=0;i<argc;i++)
            cout<<"argv["<<i<<"]="<<argv[i]<<"\n";
        myFun(argv[1],m,n);
        cout<<"****************************"<<"\n";
}
```

运行程序采用如下格式：

```
ch9-10 0
```

程序运行结果如下：

```
****************************
*    带有参数的 main 函数    *
****************************
m=1
n=2
argc=2
argv[0]=ch9-10
argv[1]=0
*b=0
c=0
最大值：2
****************************
```

分析一下程序的输出：在程序输出中有一行argc=2，这个输出表明该程序有两个参数，下面的输出分别显示了这两个参数值：

```
argv[0]=ch9-10
argv[1]=0
```

可以看到，程序的第一个参数就是程序的名字，第二个输出才是我们输入的参数0。在程序中由主函数调用自定义函数myFun时，可以改变一下参数的取得方式：

```
myFun(argv[argc-1],m,n);
```

这样可以使程序的调用更灵活。

7.9 main函数的返回值类型

其实很多C语言方面的书都没有讲清楚这个问题，甚至笔者本人都没有注意到这样的问题。你可能见过以下三种main()函数：

```
void main(){   // 返回类型为void

    …

}


main(){   // 无返回类型

    …

}


int main(){// 返回类型为整型

    …

    return 0;

}
```

可能有人有点困惑：到底main函数的返回值是什么类型呢？实际上，main函数是有返回值的（想象一下：main函数由系统调用，执行结束后，系统是不是想知道它执行的结果是什么样子，比如该如何结束），它的返回值类型是int。

千万不要写void型的main函数，在Turbo C中，如果函数没有说明返回值类型，则自动认为是int类型（虽然这对我们而言，void更直观）。对于有些编译器，如果在返回值为int的main函数中没有return语句，则该编译器也会自动在程序末尾加上return 0。而在Turbo C里，即使像下面这样的代码，也不会有错误提示，甚至连警告都没有。

```
#include <stdio.h>


int test(void){

    printf("shosh");

}


int main(void){

    printf("%d\n",test());

}
```

执行结果是输出shosh5。这个5明显是在main函数里输出的，而且刚好是test函数里的printf("shosh")的返回值。

将上面的代码修改一下，变成：

```
#include <stdio.h>
```

```
int test(void){
    …
}

int main(void){
    printf("%d\n",test());
}
```

则执行结果就是输出0。如果将int test(void)改成void test(void)则会出错，这说明编译器确实给应该有返回值而没有返回值的函数自动加了返回值，至于返回什么样的值却不可一概而论。

由此可见，第二种写法未必就存在错误（看编译器是否会帮它返回，如果返回，那会帮它返回什么样的值呢），但是肯定不是一种好习惯：一是程序不直观，容易让人误解；二是程序返回值有很大的不确定因素，在某些情况下，这可能就是错误。

比较正规的写法是将函数的返回值类型和返回语句都明确地写出来。

另外，因为Visual C++不会在生成的目标文件中加入return 0语句，所以在使用VC时要养成加入返回值语句的好习惯。

7.10　上机操作

实例7-11　利用指针函数比较数值的大小。

```
#include <iostream.h>

void inputFun();
int *compareFun(int x,int y);

void inputFun(){
    int a,b;
    int max;

    cout<<"a=";
    cin>>a;
    cout<<"b=";
    cin>>b;

    max=*compareFun(a,b);
    cout<<"Max="<<max<<"\n";
```

```
}

int *compareFun(int x,int y){
    int *p1,*p2,*p3;

    p1=&x;
    p2=&y;
    if (x<y)  {
        p3=p1;
        p1=p2;
        p2=p3;
    }
    return(p1);
}


main(){
    cout<<"****************************"<<"\n";
    cout<<"*    利用指针比较数值大小    *"<<"\n";
    cout<<"****************************"<<"\n";
    inputFun();
    cout<<"****************************"<<"\n";
}
```

程序运行结果如下：

```
****************************
*    利用指针比较数值大小    *
****************************
a=1
b=2
Max=2
****************************
```

这个例子用来比较用户输入的两个值的大小，并将其中较大的一个变量值输出。需要注意的是，在比较的过程中，并没有改变变量a和b的值，而是利用指针变量*p1和*p2分别指向这两个变量的地址，当程序比较出两个值的大小后，再将两个指针指向的地址做交换，而不改变变量本身的值。在指针交换的过程中，用到了一个指针变量*p3作为指针交换的中间变量，这样利用指针交换就达到了数值比较的目的。

實例7-12　利用指针变量比较数值大小。

```
#include <iostream.h>

void inputFun();
void sortFun(int *x, int n);

void inputFun(){
    int *p,i,a[10];

    cout<<" 请为数组输入元素值 "<<"\n";
    for(i=0;i<10;i++)  {
        cout<<"a["<<i<<"]=";
        cin>>a[i];
    }
    p=a;

    cout<<" 你输入的数组是："<<"\n";
    cout<<"a[10]={";

    for (i=0;i<10;i++)  {
        cout<<a[i]<<"";
    }
    cout<<"}"<<"\n";

    sortFun(p,10);
    cout<<" 排序后的数组："<<"\n";
    cout<<"a["<<10<<"]={";
    for(p=a,i=0;i<10;i++)  {
        cout<<*p++<<"";
    }
    cout<<"}"<<"\n";
}

/* 排序函数 */
void sortFun(int *x, int n) {
    int i, j, k, temp;
```

```
    for(i=0; i<n-1; i++) {
        k=i;
        for(j=i+1; j<n; j++)
            if(*(x+k)<*(x+j))
                k=j;
        if(k!=i) {
            temp=*(x+i); *(x+i)=*(x+k); *(x+k)=temp;
        }
    }
}
main(){
    cout<<"********************************************"<<"\n";
    cout<<"*            利用指针对数组排序            *"<<"\n";
    cout<<"********************************************"<<"\n";
    inputFun();
    cout<<"********************************************"<<"\n";
}
```

程序运行结果如下：

```
********************************************
*            利用指针对数组排序            *
********************************************
请为数组输入元素值
a[0]=1
a[1]=2
a[2]=5
a[3]=2
a[4]=9
a[5]=0
a[6]=10
a[7]=98
a[8]=99
a[9]=1000
你输入的数组是：
a[10]={1 2 5 2 9 0 10 98 99 1000 }
排序后的数组：
```

```
a[10]={1000 99 98 10 9 5 2 2 1 0 }
**************************************
```

请根据程序的输出结果，自行对程序做出分析。

7.11 小结

本章讲述的指针也是C++的一种数据类型，并且是C++中非常重要的数据类型。指针使得C++程序更加灵活、简洁、紧凑和高效。特别对于系统软件来说，指针能够解决一些使用通常方法不能解决的问题。在学习指针时，应当着重理解相关的概念：指针、指针变量、地址、取地址、指针引用、指针数组与指向数组的指针、指针运算规则等。

指针在给程序带来巨大方便的同时也会有其副作用，关于指针的知识、使用方法等都非常复杂，读者在学习和使用指针时一定要认真、仔细。

7.12 习题

一、填空题

1．变量地址就是_____。

2．变量的访问方法有两种，分别是：_____和_____。

3．一个指针在被定义后应该被赋予相应的值，也就是要让它指向一个地址，否则指针就是_____，而_____对于系统是有害的。

4．指针变量只能存放_____，即内存地址。

5．一个指针变量只能存放_____变量的地址。

二、程序分析题

认真阅读下面的程序，手工写出程序的运行结果，并上机验证。

1．程序代码如下：

```cpp
#include <iostream.h>


main(){
    int a[10];
    int *p;
    int i;


    for (i=0;i<10;i++)  {
        cout<<"a["<<i<<"]=";
        cin>>a[i];
```

C++语言设计教程

```
    }

    p=a;
    cout<<"a[10]={";
    for (i=9;i>=0;i--)  {
        cout<<*p++;
        if (i>0)
            cout<<",";
    }
    cout<<"}"<<"\n";
}
```

2. 程序代码如下：

```
#include <iostream.h>

main(){
    int a,b;
    int *p;

    cout<<"a=";
    cin>>a;
    cout<<"b=";
    cin>>b;

    p=&a;
    cout<<*p<<"->";
    *p=a+1;
    p=&b;
    cout<<*p<<"->";
    *p=b+2;cout<<a<<"->"<<b<<"\n";
}
```

3. 程序代码如下：

```
#include <iostream.h>

const int N=10;

main(){
```

```
        int a[N];
        int max,min;
        int *p;
        int i;

        for (i=0;i<N;i++)  {
            cout<<"a["<<i<<"]=";
            cin>>a[i];
        }

        cout<<"a["<<N<<"]={";
        for (i=0;i<N;i++)  {
            cout<<a[i];
            if (i<N-1)
                cout<<",";
        }
        cout<<"}";
        cout<<"\n";

        max=*a;
        min=*a;
        for (p=a+1;p<a+N;p++)  {
            if (*p>max)
                max=*p;
            else if (*p<min)
                min=*p;
        }

        cout<<"Max="<<max<<"\n";
        cout<<"Min="<<min<<"\n";
}
```

4. 程序代码如下：

```
#include <iostream.h>


main(){
    int a[10];
```

```
        int *p=a;
        int i;

        for (i=0;i<10;i++)  {
              cout<<"a["<<i<<"]=";
              cin>>a[i];
        }

        cout<<"*p++:";
        for (i=0;i<10;i++)
              cout<<""<<*p++;
        cout<<"\n";

        cout<<"++*p:";
        for (i=0;i<10;i++)
              cout<<""<<++*p;
        cout<<"\n";

        cout<<"*--p:";
        for (i=0;i<10;i++)
              cout<<""<<*--p;
        cout<<"\n";

        cout<<"--*p:";
        for (i=0;i<10;i++)
              cout<<""<<--*p;
        cout<<"\n";
}
```

5. 程序代码如下：

```
#include <iostream.h>

main(){
    int a[2][3]={{1,2,3},{4,5,6}};
    int *p;
    int i,j;
```

```
p=&a[0];

for (i=0;i<2;i++) {
    for (j=0;j<3;j++) {
        cout<<""<<*p++;
    }
    cout<<"\n";
}
cout<<"\n";
}
```

三、问答题

1. 什么是指针?

2. 为什么要使用指针?

3. main()函数是否可以有形式参数? 其一般形式是什么?

4. main()函数的返回值类型是什么?

第8章 结构体、链表与联合

本章学习目标

▶ 结构体的概念与应用   ▶ 联合的概念与应用

▶ 链表的概念   ▶ 链表的操作

前面已经讲述了C++的一些重要的数据类型，例如：整型、字符型、浮点型、数组和指针。这些数据类型无一例外都是单一的数据类型，数组虽然由若干数组元素组成，但是数组元素都必须拥有相同的数据类型，而且数组的数据类型与数组元素的数据类型是一致的。在实际生产、生活中，仅有这些数据类型是不够的，举个例子：一个学生的详细信息，包括姓名、年龄、性别以及各科成绩等，这些信息是一个有机的整体，一般不希望用不同的变量来表示，因为这样会显得杂乱无章，如何解决呢？C++提供了另外一种数据类型，它能包含多种数据类型，这就是本章要讲的结构体。

链表是C++的重要概念。前面章节中讲过C++的常用数据类型，而链表则是C++的另外一种重要的数据类型。前面讲到的数组在定义时必须指定数组元素的数量，这样编译程序在为数组分配空间时才可以根据数组元素的数量来分配空间。在实际情况中，很有可能不知道数组有多少元素，或者数组元素的数量可能有变化，这时就必须定义一个足够大的数组，以满足数组元素变化的要求。但是这样会造成空间上的浪费，而且庞大的数组对系统的效率也是有害的。那么能不能动态分配空间呢？答案是肯定的。

本章讲到的链表就是动态数组，它是在程序运行过程中，根据需要动态开辟内存空间，这样既能有效地节约空间，又能有效地提高程序的执行效率。

链表是一种复杂的数据结构，根据数据之间的关系可分为：单链表、循环链表和双向链表3种。

8.1 结构体的概念

简单地讲，结构体就是把许多数据类型不相同的变量组织起来，成为一种新的数据类型，这就是结构体。结构体拥有不同数据类型的成员，是不同数据类型成员的集合。

> **提示**
>
> 数组是把具有相同数据类型的数据按照一定的次序组成的集合，结构体是把具有不同数据类型的数据按照一定的次序组成的集合。

8.2 结构体的定义和使用

结构体可以拥有不同数据类型的成员。

在C++中定义结构体的一般形式如下：

```
struct 结构体名{
    数据类型 成员1;
    数据类型 成员2;
    …
    数据类型 成员n;
}
```

例如：一个学生的详细信息，包括学号、姓名、性别、年龄、语文成绩、数学成绩和英语成绩等。针对学生信息这个例子，可以定义如下的结构体来表示：

```
struct studentInfo{
    char NO[10]; /* 学号 */
    char Name[10]; /* 姓名 */
    char Sex[2]; /* 性别 */
    short int Age; /* 年龄 */
    float Chinese; /* 语文成绩 */
    float Math; /* 数学成绩 */
    float English; /* 英语成绩 */
}
```

结构体定义后，需要再定义一个结构体类型的变量，以便在程序中使用这个结构体。结构体变量的定义可以分为3种方式。

1. 普通定义方式

这种定义方式与普通变量的定义方式相同，只是在定义结构体变量之前，必须要先定义结构体。例如：

```
struct studentInfo{
    char NO[10]; /* 学号 */
    char Name[10]; /* 姓名 */
    char Sex[2]; /* 性别 */
    short int Age; /* 年龄 */
    float Chinese; /* 语文成绩 */
    float Math; /* 数学成绩 */
    float English; /* 英语成绩 */
}
struct studentInfo student1,student2;
```

在上例中，先定义了一个结构体studentInfo，然后再用这个结构体定义了两个结构体类型的变量student1和student2。需要注意的是，在使用这种方法定义结构体类型的变量时，不仅需要指定该变量的类型是结构体类型，而且还需要指定是哪一个结构体类型。

2．定义结构体类型的同时定义结构体变量

这种定义方式是在定义结构体时就定义结构体变量，同时这个结构体也可以再定义其他结构体变量。例如：

```
struct studentInfo{
    char NO[10]; /*学号*/
    char Name[10]; /*姓名*/
    char Sex[2]; /*性别*/
    short int Age; /*年龄*/
    float Chinese; /*语文成绩*/
    float Math; /*数学成绩*/
    float English; /*英语成绩*/
} student1,student2;
```

在这个例子中，在定义结构体studentInfo的同时又定义了两个该结构体类型的变量student1和student2。

3．直接定义结构体变量而不定义结构体名称

这种方式比较少见，它是在定义结构体时不给出结构体的名字，而是直接定义结构体的变量。这种定义方式与第二种定义方式的不同之处在于，后者在定义完成之后，还可以再定义其他该结构体类型的变量，但是这种方式由于没有结构体名称，所以不能再定义该结构体类型的变量。例如：

```
struct{
    char NO[10]; /*学号*/
    char Name[10]; /*姓名*/
    char Sex[2]; /*性别*/
    short int Age; /*年龄*/
    float Chinese; /*语文成绩*/
    float Math; /*数学成绩*/
    float English; /*英语成绩*/
} student1,student2;
```

需要注意的是，在一个程序中，普通变量的名字可以和结构体中的成员变量的名字相同，它们代表不同的含义。例如：

```
struct studentInfo{
    char NO[10]; /*学号*/
    char Name[10]; /*姓名*/
```

```
  char Sex[2]; /* 性别 */
  short int Age; /* 年龄 */
  float Chinese; /* 语文成绩 */
  float Math; /* 数学成绩 */
  float English; /* 英语成绩 */
} student1,student2;
…
short int Age;
```

在这个例子中，有一个普通的short int类型的变量Age，在结构体studentInfo中也有一个同样的变量，它们是完全不同的两个变量，互不干扰。

8.3 使用结构体变量

在8.2节中讲述了如何定义一个结构体，以及如何定义结构体变量。在定义完成后，接下来就要使用结构体。使用结构体是通过结构体变量来实现的。

在C++中，结构体变量的引用格式如下：

```
结构体变量 . 结构体成员；
```

其中的.号称为成员运算符，它在所有运算符中拥有最高的等级。例如，定义了一个结构体，并且定义了一个该结构体的变量：

```
struct studentInfo{
  char NO[10]; /* 学号 */
  char Name[10]; /* 姓名 */
  char Sex[2]; /* 性别 */
  short int Age; /* 年龄 */
  float Chinese; /* 语文成绩 */
  float Math; /* 数学成绩 */
  float English; /* 英语成绩 */
} student;
```

这时，可以使用下面的方式引用该结构体中的各成员：

```
student.NO;
student.Name;
student.Sex;
student.Math;
```

需要注意的是，在引用结构体时是引用结构体的成员，而不能直接引用结构体本身，例如：

```
student.studentInfo;
```

这种引用方式是错误的。

8.4 结构体的初始化

完成结构体定义后便可以对它进行初始化，因为结构体中包含不同类型的变量，所以对它做初始化时要根据各成员数据类型的不同赋予不同类型的值。例如：

```
struct studentInfo{
    char NO[10]; /* 学号 */
    char Name[10]; /* 姓名 */
    char Sex[2]; /* 性别 */
    short int Age; /* 年龄 */
    float Chinese; /* 语文成绩 */
    float Math; /* 数学成绩 */
    float English; /* 英语成绩 */
}
struct studentInfo student{"0000000001","ShuLi","M",23,99.5,100,99};
```

完成结构体的初始化后就可以按照上一节讲的结构体变量的引用方式来引用这些变量了。下面通过一个例子来演示如何为结构体赋值并输出各成员。

实例8-1　结构体初始化。

```
#include <iostream.h>

void strucFun();

void strucFun(){
    struct studentInfo{
        char NO[10]; /* 学号 */
        char Name[10]; /* 姓名 */
        char Sex[2]; /* 性别 */
        short int Age; /* 年龄 */
        float Chinese; /* 语文成绩 */
        float Math; /* 数学成绩 */
        float English; /* 英语成绩 */
    };
    struct studentInfo student={"000000001","ShuLi","M",23,99.5,100,99};

    cout<<"学号："<<student.NO<<"\n";
    cout<<"姓名："<<student.Name<<"\n";
    cout<<"性别："<<student.Sex<<"\n";
```

```
    cout<<" 年龄："<<student.Age<<"\n";

    cout<<" 语文成绩："<<student.Chinese<<"\n";

    cout<<" 数学成绩："<<student.Math<<"\n";

    cout<<" 英语成绩："<<student.English<<"\n";

    cout<<"******************"<<"\n";

    cout<<" 总成绩："<<student.Chinese+student.Math+student.English<<"\n";

    cout<<" 平均成绩："<<(student.Chinese+student.Math+student.English)/3<<"\n";
}


main(){
    cout<<"***********************************************************"<<"\n";
    cout<<"*                    结构体的初始化                        *"<<"\n";
    cout<<"***********************************************************"<<"\n";
    strucFun();
    cout<<"***********************************************************"<<"\n";
}
```

程序运行结果如下：

```
***********************************************************
*                    结构体的初始化                        *
***********************************************************
学号：000000001
姓名：ShuLi
性别：M
年龄：23
语文成绩：99.5
数学成绩：100
英语成绩：99
******************
总成绩：298.5
平均成绩：99.5
***********************************************************
```

8.5　结构体数组

在前面几节介绍了结构体的概念和基本使用方法。在解决实际问题时，用到结构体最多的地方是结构体数组。还是以学生信息为例，如果希望了解学生的整体信息，使用结构体数组比使用单个变量更具有整体性。

结构体数组的一般形式如下：

```
struct 结构体名{
    数据类型 成员1;
    数据类型 成员2;
    …
    数据类型 成员n;
} 变量名列表 [n]
```

结构体数组的引用方式为：

```
变量名 [n]. 成员变量
```

实例8-2 结构体数组的应用。

```cpp
#include <iostream.h>


void strucFun();


void strucFun(){
    struct studentInfo{
        char NO[10]; /* 学号 */
        char Name[10]; /* 姓名 */
        char Sex[2]; /* 性别 */
        short int Age; /* 年龄 */
        float Chinese; /* 语文成绩 */
        float Math; /* 数学成绩 */
        float English; /* 英语成绩 */
    }student[2]={{"000000001","ShuLi","W",23,99.5,100,99},{"000000002",
"LiHong","M",25,100,90,80}};


    cout<<"=== 学生 :"<<student[0].Name<<" 的信息 ==="<<"\n";
    cout<<" 学号 :"<<student[0].NO<<"\n";
    cout<<" 姓名 :"<<student[0].Name<<"\n";
    cout<<" 性别 :"<<student[0].Sex<<"\n";
    cout<<" 年龄 :"<<student[0].Age<<"\n";
    cout<<" 语文成绩 :"<<student[0].Chinese<<"\n";
    cout<<" 数学成绩 :"<<student[0].Math<<"\n";
    cout<<" 英语成绩 :"<<student[0].English<<"\n";
```

```
    cout<<"******TOTAL******"<<"\n";
    cout<<" 总成绩 :"<<student[0].Chinese+student[0].Math+student[0].English<<"\n";
    cout<<" 平 均 成 绩 :"<<(student[0].Chinese+student[0].Math+student[0].
English)/3<<"\n";
    cout<<"*****************"<<"\n";


    cout<<"=== 学生 :"<<student[1].Name<<" 的信息 ==="<<"\n";
    cout<<" 学号 :"<<student[1].NO<<"\n";
    cout<<" 姓名 :"<<student[1].Name<<"\n";
    cout<<" 性别 :"<<student[1].Sex<<"\n";
    cout<<" 年龄 :"<<student[1].Age<<"\n";
    cout<<" 语文成绩 :"<<student[1].Chinese<<"\n";
    cout<<" 数学成绩 :"<<student[1].Math<<"\n";
    cout<<" 英语成绩 :"<<student[1].English<<"\n";
    cout<<"******TOTAL******"<<"\n";
    cout<<" 总成绩 :"<<student[1].Chinese+student[1].Math+student[1].English<<"\n";
    cout<<" 平 均 成 绩 :"<<(student[1].Chinese+student[1].Math+student[1].
English)/3<<"\n";
    cout<<"*****************"<<"\n";
}


main(){
    cout<<"*************************************************************"<<"\n";
    cout<<"*                    结构体数组                           *"<<"\n";
    cout<<"*************************************************************"<<"\n";
    strucFun();
    cout<<"*************************************************************"<<"\n";
}
```

程序运行结果如下：

```
*************************************************************
*                    结构体数组                           *
*************************************************************
=== 学生 :ShuLi 的信息 ===
学号 :000000001
姓名 :ShuLi
性别 :W
```

```
年龄：23
语文成绩：99.5
数学成绩：100
英语成绩：99
******TOTAL******
总成绩：298.5
平均成绩：99.5
****************
=== 学生：LiHong 的信息 ===
学号：000000002
姓名：LiHong
性别：M
年龄：25
语文成绩：100
数学成绩：90
英语成绩：80
******TOTAL******
总成绩：270
平均成绩：90
****************
************************************************************
```

8.6 链表的概念

  按照链表的字面理解，可以给它下一个通俗易懂的定义：若干个数据项按照一定的规则排列并通过指针链接起来，每个数据项都包含一个指向下一个数据项的指针，最后一个数据项没有指向固定目标的指针，其值为NULL，这些链接到一起的数据项称为链表。

  数组中的元素在内存中是连续存储的，但是链表不需要，因为链表中的每个数据（节点）都包含一个指向下一个数据的指针，根据这个指针可以很方便快捷地找到下一个数据，直到这个链表的末端。要在链表中找到一个节点，必须找到这个节点的上一个节点，按照这种方法可以从一个链表的第一个节点顺序往下找，直到找出这个链表的所有节点为止。

8.7 节点

  在8.6节描述了链表的概念，其中一个重要的名词就是数据项。在链表中，每个数据项就是一个节点。链表中的每个节点都包括两部分：数据和指针。数据是用户真正需要的内容，而指针用来指向下一个节点。图8-1是一个典型的单链表结构图，它描述了单链表的完整结构。

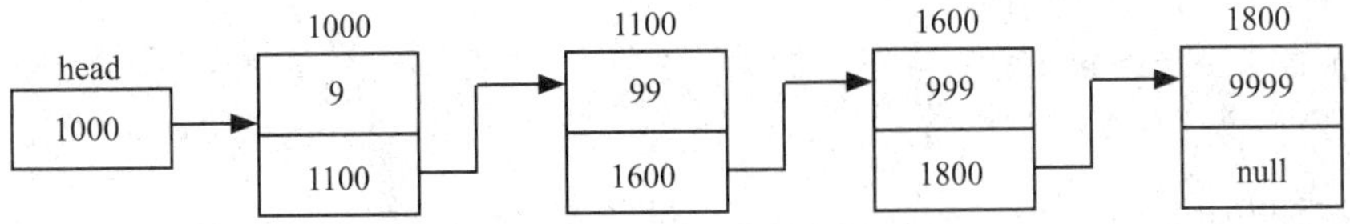

图8-1　单链表的结构

每个链表在内存中都必须有一个头（head），这个是链表在内存中的首地址。除首节点外，每个节点都包括数据和指针两部分。在图8-1中，首节点的指针指向1000，代表它的下一个节点的地址是1000，通过这个指针可以找到第二个节点（1000），第二个节点有两个部分组成：9和1100，其中9是数据部分，是用户真正需要的数据，而1100是第三个节点的地址，通过这个地址可以找到第三个节点（1100）。依此类推，能够找到这个链表的所有节点，也就能访问到每个节点的数据，并找到这个链表的最后一个节点。链表的最后一个节点包含了数据部分和一个空指针（NULL），因为这个节点的指针不再指向一个具体的目标，所以为空指针。

从图8-1可以看出，要找到链表中的每个数据都必须从链表头开始顺序往后搜索，直到找到需要的数据为止。另外，从图中还可以看出，链表的每个节点的地址不连续，例如第三个节点的地址是1100，第四个节点的地址变成了1600。因为每个节点都有指向下一个节点的指针，所以整张链表能够链接到一起，不会出现断链。

节点的数据结构如下：

```
struct node{
  int a;
  char b;
  char c[];
  …
  struct node *p;
};
```

在这个数据结构中：

```
  int a;
  char b;
  char c[];
  …
```

表示是数据部分，而接下来的部分：

```
  struct node *p;
```

表示是指针部分。

8.8　创建单链表

要使用链表就必须首先创建链表，而创建链表必须先创建链表节点。下面通过一个例子来说明链表创建和输出的方法。

实例8-3　创建并输出链表。

```cpp
#include <iostream.h>


// 定义链表节点结构
struct Node{
    int data;
    Node *next ;
};


// 创建链表，要求用户输入数据，直到输入 -9999。
void createLink( Node *&list){
    int num;
    Node *p  ;
    cout<<" 请输入链表中节点的数据（-9999 表示结束）"<<"\n";
    cin>>num;
    while(num!=-9999)  {
        p = new Node;
        p->data =num;
        p->next =list;
        list=p;
        cin>>num;
    }
}


// 输出链表中的数据
void outputLink( Node  *&list){
    cout<<" 链表中的数据如下 : "<<"\n";
    Node *p=list;

    while(p!=NULL){
        cout<<""<<p->data;
        p=p->next;
    }
    cout<<"\n";
}
```

```
// 主函数
int main(void){
    Node * head=NULL;
    cout<<"*****************************************"<<"\n";
    cout<<"*              创建并输出链表              *"<<"\n";
    cout<<"*****************************************"<<"\n";
    createLink(head);
    outputLink(head);
    cout<<"*****************************************"<<"\n";
}
```

程序运行结果如下：

```
*****************************************
*              创建并输出链表              *
*****************************************
请输入链表中节点的数据（-9999 表示结束）
9
99
999
9999
99990
-9999

链表中的数据如下：
 99990 9999 999 99 9
*****************************************
```

在这个例子中，首先定义了链表的节点Node，每个节点就是一个Node，每个Node包含一个整型数据和一个指针。

接下来定义了一个自定义函数createLink，使用这个函数时要求用户输入数据，并使用这些数据来创建链表。在这个函数中还定义了一个整型值num，用来接收用户输入的数据。还定义了一个Node类型的指针，用来将用户输入的值num写进链表中，并形成指向下一个节点的指针。

再接下来定义了一个自定义函数outputLink，使用这个函数输出刚才用createLink函数创建的链表。在这个函数中，同样定义了一个Node类型的指针，用来逐个输出链表节点中的数据部分，直到链表中的指针为NULL为止，表示整个链表中的数据部分已经全部输出完毕。

在主函数中，定义一个Node类型的指针head，并将这个指针的值置为NULL，然后逐个调用函数createLink和outputLink来创建和输出链表。

8.9 向链表插入节点

链表是动态存储数据的，只有在需要的时候才向系统申请空间，用于保存数据。这就涉及向链表中插入节点的问题。向链表中插入节点可分为3种情况：

(1) 向链表头插入节点。

(2) 向链表尾插入节点。

(3) 向链表中间插入节点。

图8-2至图8-4分别表示这3种插入节点的方法，图中虚线表示插入新的节点后指针的指向。向链表头插入节点，实现语句为：

```
head=p1;
```

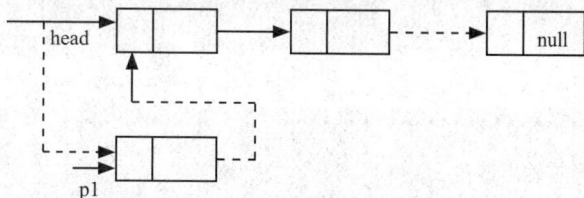

图8-2　在链表头插入节点

这种情况是向原来完整的链表头部插入新节点，使链表的 head 指向新插入的节点。head 指向新插入的节点 p1，由节点 p1 再指向原来链表的节点。整个插入过程不会破坏原来链表的完整性。

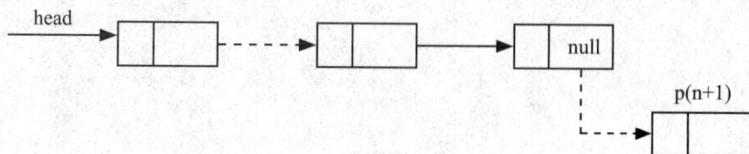

图8-3　在链表尾插入节点

这种插入方法是在链表尾插入节点。因为链表尾部的指针没有指向任何明确的目标，它的值是null，所以在它后面加上新节点就需要把原来尾部的指针指向新加入的节点，并把新加入节点中的最后一个节点的指针的值置为null值。实现语句为：

```
p(n)->next=p(n+1);

p(n+1)->next=null;
```

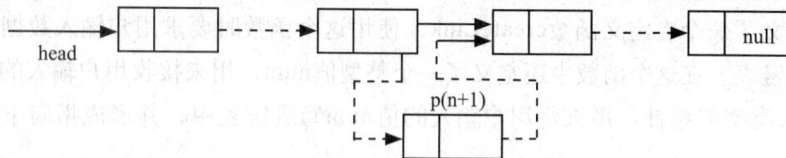

图8-4　在链表中间插入节点

这种插入方法是在链表中间某一位置插入节点。由于链表中间的任一节点都被它前面的节点所指向，同时它自己又指向其后面的节点，因此，在插入新节点时，需要先将新插入节点的指针指向原先插入点后面的节点，再将插入点前面节点的指针指向新插入的节点，实现语句为：

```
p(n+1)->next=p(n)->next;
```

```
p(n)->next=p(n+1);
```

实例8-4 链表的建立、插入和输出。

```
#include <iostream.h>

struct myNode{
    int data;
    myNode *next ;
};


/* 建立链表 */
void createMyLink( myNode *&list){
    int data;
    myNode    *Node  ;
    int i;

    for (i=0;i<10;i++)  {
        Node = new myNode;
        Node->data =i;
        Node->next =list;
        list=Node;
    }
}


/* 输出链表 */
void outputMyLink( myNode  *&list){
    cout<<" 本链表结构如下：" <<"\n";
    myNode *p=list;

    while(p!=NULL){
        cout<<" "<<p->data;
        p=p->next;
    }
    cout<<"\n";
}


/* 在链表中插入节点 */
```

```
/*注意：这里有两个变量data*/
void insertMyNode(myNode *  &list){
    int data,tempNode;
    int i;
    myNode *p = list;
    myNode *newNode = new myNode;

    cout<<"在第2个节点插入值99"<<"\n";
    tempNode=2;
    data=99;
    newNode->data = data;
    if(tempNode == 0||tempNode == 1)     {
        newNode->next = p;
        list = newNode;
        return;
    }

    i = 1;
    while(i++<tempNode-1 && p->next!=NULL )
        p = p->next;
    if(p->next!=NULL)     {
        newNode->next = p->next;
        p->next = newNode;
        return;
    }
    p->next = newNode;
    newNode->next = NULL;
}


main(){
    cout<<"*******************************************"<<"\n";
    cout<<"*            链表的建立、插入和输出            *"<<"\n";
    cout<<"*******************************************"<<"\n";
    /*定义头指针，并将其值赋为null。*/
    myNode * head=NULL;
    createMyLink(head);
```

```
        outputMyLink(head);

        insertMyNode(head);

        outputMyLink(head);

        cout<<"**********************************************"<<"\n";

}
```

程序运行结果如下：

```
**********************************************

*              链表的建立、插入和输出          *

**********************************************

本链表结构如下：

 9 8 7 6 5 4 3 2 1 0

在第 2 个节点插入值 99

本链表结构如下：

 9 99 8 7 6 5 4 3 2 1 0

**********************************************
```

在这个例子中，把链表的建立、新节点的插入和链表的输出分别作为独立的功能函数来实现。在函数createMyLink中，通过循环语句建立了一个简单的链表；接着用显示函数outputMyLink中将其输出，再通过insertMyNode函数在这个已建立的链表的第二个节点处插入一个值为99的节点。

8.10　删除链表中的节点

类似于链表中节点的插入，从链表中删除节点也有3种情况，分别是：

（1）删除链表头节点。

（2）删除链表尾节点。

（3）删除链表中间的某一个节点。

图8-5至图8-7分别描述了删除链表中节点的3种情况。

删除链表头节点的实现语句如下：

```
head=head->next;
```

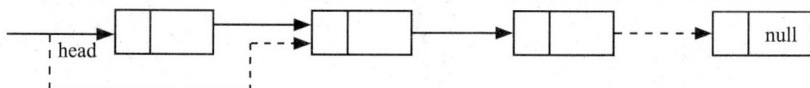

图8-5　删除链表头节点

在一个链表中，指针head直接指向头节点。当头节点被删除时，必须使head指向头节点后面的第二个节点，以维持整个链表的完整性。

删除链表的尾节点的实现语句如下：

```
p->next=null;
```

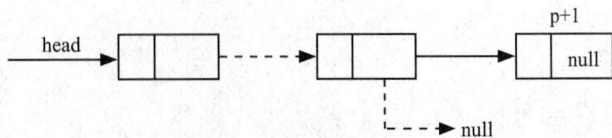

图8-6 删除链表尾节点

链表的尾节点含有一个值为null的指针，如果要删除这个尾节点，则必须把位于尾节点前面的一个节点的指针值变为null。

删除链表中间任一节点的实现语句如下：

```
p(n)->next=p(n+1)->next;
```

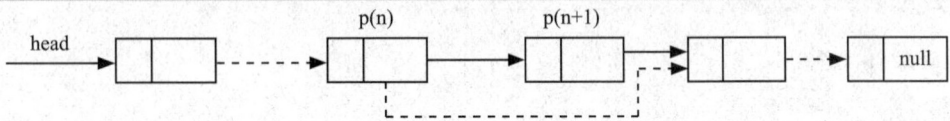

图8-7 删除链表中间节点

链表中间的任意一个节点都是它前面节点指向的目标，同时它自己又指向其后面的节点。如果要删除中间一个节点，则必须把它前面的节点的指针指向它后面的节点，以保证整个链表的完整性。

实例8-5 删除链表的节点。

```cpp
#include <iostream.h>

struct myNode{
    int data;
    myNode *next ;
};

/* 建立链表 */
void createMyLink( myNode *&list){
    int data;
    myNode    *Node  ;
    int i;

    for (i=0;i<10;i++)    {
        Node = new myNode;
        Node->data =i;
        Node->next =list;
        list=Node;
    }
}
```

```
/* 输出链表 */
void outputMyLink( myNode  *&list){
    cout<<" 本链表结构如下 : "<<"\n";
    myNode *p=list;

    while(p!=NULL){
        cout<<""<<p->data;
        p=p->next;
    }

    cout<<"\n";
}


/* 删除节点 */
void deleteMyLink( myNode *&list){
    int tempNode;
    int i;
    myNode *p = list;

    cout<<" 下面要删除第 2 个节点 : "<<"\n";
    tempNode=2;

     i = 2;
    while(i<tempNode && p!=NULL ) {
        p = p->next;
        i++;
    }

    if(p!=NULL) {
        p->next = p->next->next;
    } else {
        cout<<" 没有这个节点 ";
    }
}
```

```
main(){
    cout<<"*******************************************"<<"\n";
    cout<<"*              删除链表的节点              *"<<"\n";
    cout<<"*******************************************"<<"\n";

    /*定义头指针，并将其值赋为 null。*/
    myNode *head=NULL;

    createMyLink(head);
    outputMyLink(head);
    deleteMyLink(head);
    outputMyLink(head);
    cout<<"*******************************************"<<"\n";
}
```

程序运行结果如下：

```
*******************************************
*              删除链表的节点              *
*******************************************
本链表结构如下：
 9 8 7 6 5 4 3 2 1 0
下面要删除第 2 个节点：
本链表结构如下：
 9 7 6 5 4 3 2 1 0
*******************************************
```

8.11 联合的概念

与结构体一样，联合也是C++中的用户自定义数据类型。联合的定义与结构体一样，也是由不同的数据成员构成，这些数据成员也可以拥有不同的数据类型，甚至可以有自己的成员函数。

但是，联合与结构体有一点不同：任一时刻，结构体中的任何成员都可以被访问，而联合中的成员函数却只有一个可以被访问，其余的成员则是不可以被访问的。

结构体和联合的不同点反映在内存分配上是这样的：结构体中的每个成员都有自己独立的内存空间分配，而整个结构体的内存空间就是该结构体中所有成员所分配的内存空间之和。

联合中的成员没有分配自己独立的内存空间，每一时刻都可能是一个不同的成员在使用该联合的内存，同时其他成员也就不能再使用该联合的内存。联合的内存空间的大小是由该联合的成员中占有内存空间最大的一个成员所占有的内存空间来决定的。

联合的定义如下：

```
union 联合名{
    数据类型 成员1;
    数据类型 成员2;
    …
    数据类型 成员n;
}变量列表；
```

例如：

```
union data{
int a;
float b;
double c;
char d ;
}a;
```

下面定义一个结构体和一个联合，对比一下它们占用内存空间的情况。

结构体的定义代码如下：

```
struct data {
    int a;
    float b;
    double c;
    char d ;
};
```

联合的定义代码如下：

```
struct data{
    int a;
    float b;
    double c;
    char d ;
};
```

图8-8表示了上面的结构体和联合占用内存的情况。

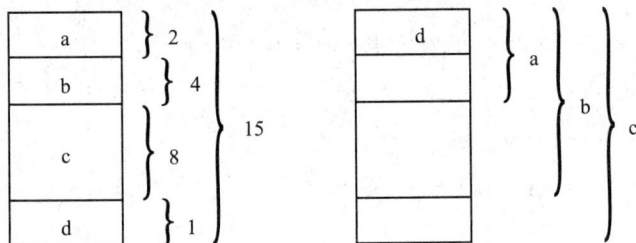

图8-8　结构体和联合占用内存对比

可以看出，上面定义的结构体和联合所包含的成员的数据类型是一样的，结构体中的每个成员都拥有自己独立的内存空间，而联合中的成员则是共享同一段内存空间。反映到整体内存占用上，这个结构体占用的内存是15字节，而与它成员相同的联合却只占用了8字节。很明显，联合比结构体能够节省内存空间。

可以用下面的例子来验证图8-8。

实例8-6　结构体和联合占用内存的情况。

```cpp
#include <iostream.h>

/* 定义结构体 */
struct data1{
    short int a;
    float b;
    double c;
    char d;
};


/* 定义拥有相同数据成员的联合 */
union data2{
    short int a;
    float b;
    double c;
    char d;
};


main(){
    struct data1 sd;
    union data2 ud;
    cout<<"*****************************************"<<"\n";
    cout<<"*          结构体和联合占用内存比较          *"<<"\n";
    cout<<"*****************************************"<<"\n";
    cout<<"=== 结构体占用的内存空间 ==="<<"\n";
    cout<<"a:"<<sizeof(sd.a)<<"\n";
    cout<<"b:"<<sizeof(sd.b)<<"\n";
    cout<<"c:"<<sizeof(sd.c)<<"\n";
    cout<<"d:"<<sizeof(sd.d)<<"\n"";
    cout<<" 总空间 :"<<sizeof(sd)<<"\n";
```

```
        cout<<"=== 联合占用的内存空间 ==="<<"\n";
        cout<<"a:"<<sizeof(ud.a)<<"\n";
        cout<<"b:"<<sizeof(ud.b)<<"\n";
        cout<<"c:"<<sizeof(ud.c)<<"\n";
        cout<<"d:"<<sizeof(ud.d)<<"\n";
        cout<<" 总空间 :"<<sizeof(ud)<<"\n";
        cout<<"*******************************"<<"\n";
}
```

程序的运行结果如下：

```
*******************************
"*        结构体和联合占用内存比较        *"
"*******************************"
=== 结构体占用的内存空间 ===
a:2
b:4
c:8
d:1
总空间 :15
=== 联合占用的内存空间 ===
a:2
b:4
c:8
d:1
总空间 :8
*******************************
```

需要说明的是，程序中的sizeof()运算符是C++提供的数学函数，使用它能够很方便地计算出每种数据类型的长度。

8.12 引用联合变量

引用联合变量与引用结构体变量形式相似，但由于联合在同一时刻只能引用其中的一个变量，所以在引用联合变量时要特别小心。

实例8-7　引用联合变量。

```
#include <iostream.h>


void myFun();
```

```
/* 定义联合 */
union data{
    int a;
    float b;
    char c;
};

void myFun(){
    union data x;

    /* 正确的引用方式 */
    x.a=10;
    cout<<"a="<<x.a<<"\n";
    x.b=1.0;
    cout<<"b="<<x.b<<"\n";
    x.c='x';
    cout<<"c="<<x.c<<"\n";
    /* 错误的引用方式 */
    x.a=100;
    x.c='y';
    cout<<"a="<<x.a<<"\n";
    cout<<"c="<<x.c<<"\n";
}

main(){
    cout<<"**************************"<<"\n";
    cout<<"*        联合变量的引用        *"<<"\n";
    cout<<"**************************"<<"\n";
    myFun();
    cout<<"**************************"<<"\n";
}
```

程序运行结果如下：

```
**************************
*        联合变量的引用        *
**************************
```

```
a=10
b=1
c=x
a=121
c=y
***************************
```

通过这个例子可以看出，逐个地引用联合成员能够得出正确的结果。但是，当同时引用两个联合成员时，第一个被引用的成员将被覆盖，从而得出一个莫名其妙的值，事实上这个值也没有任何意义。

8.13 上机操作

利用结构体录入及输出学生信息

本例将利用一个结构体来实现简单的信息录入和输出功能。本例中把录入和输出过程分别用不同的函数来实现，使程序的各个功能模块的作用非常清晰，也更容易理解和阅读。

实例8-8 学生信息管理。

```cpp
#include <iostream.h>
#include <string.h>

/* 定义结构体 */
struct stuInfo{
    char NO[10];
    char Name[10];
    float Chinese[10];
    float Math[10];
    float English[10];
};

/* 输入的过程 */
void inputBasicInfo(int n){
    struct stuInfo stu[n];
    int i;

    cout<<"===== 录入过程 ====="<<endl;
    for (i=0;i<n;i++)  {
```

```
            cout<<"学    号:";
            cin>>stu[i].NO;
            cout<<"姓    名:";
            cin>>stu[i].Name;
            cout<<"语文成绩:";
            cin>>stu[i].Chinese[i];
            cout<<"数学成绩:";
            cin>>stu[i].Math[i];
            cout<<"英语成绩:";
            cin>>stu[i].English[i];
            cout<<endl<<endl;
        }
    }


/* 输出的过程 */
void outputBasicInfo(int n){
    struct stuInfo stu[n];
    int i;

    cout<<"===== 输出过程 ====="<<endl;
    for (i=0;i<n;i++)  {
        cout<<"学    号:"<<stu[i].NO<<endl;
        cout<<"姓    名:"<<stu[i].Name<<endl;
        cout<<"语文成绩:"<<stu[i].Chinese[i]<<endl;
        cout<<"数学成绩:"<<stu[i].Math[i]<<endl;
        cout<<"英语成绩:"<<stu[i].English[i]<<endl;
        cout<<endl<<endl;
    }
}


/* 主函数 */
main(){
    int n;
    cout<<"********************************"<<endl;
    cout<<"*            学生信息           *"<<endl;
    cout<<"********************************"<<endl;
```

```
    cout<<" 请输入学生数 (n<=10) : ";

    cin>>n;

    cout<<endl<<endl;

    inputBasicInfo(n);

    outputBasicInfo(n);

    cout<<"**********************************"<<endl;

}
```

程序运行结果如下:

```
**********************************
*              学生信息            *
**********************************
请输入学生数 (n<=10) : 2


===== 录入过程 =====
学    号:001
姓    名:ShuLi
语文成绩:100
数学成绩:99.5
英语成绩:98



学    号:002
姓    名:JingJing
语文成绩:100
数学成绩:98
英语成绩:99.5



===== 输出过程 =====
学    号:001
姓    名:ShuLi
语文成绩:100
数学成绩:99.5
英语成绩:98
```

```
学    号：002
姓    名：JingJing
语文成绩：100
数学成绩：98
英语成绩：99.5


**********************************
```

在这个程序中，首先定义了一个结构体stuInfo，该结构体的成员包括学号、姓名、语文成绩、数学成绩和英语成绩。接下来定义了以下几个函数：

- inputBasicInfo函数 提供一个人机界面，方便用户录入信息，并将用户录入的信息利用结构体stuInfo保存起来。
- outputBasicInfo函数 将使用inputBasicInfo函数保存的用户录入信息按照规定的顺序读取出来。
- 主函数 负责调用inputBaiscInfo和outputBasicInfo函数。

本例还可以定义其他一些功能函数来实现更多的功能，比如求总成绩、求平均成绩等，请读者根据自己的兴趣来完成。

8.14 小结

本章讲述了结构体、链表和联合的概念。在前面讲数组时曾讲到，数组是一个具有相同数据类型的数据序列，而本章讲的结构体却是具有不同数据类型的数据序列。结构体中的变量具有不同的数据类型，它们可以是简单数据类型，也可以是复合数据类型。

联合与结构体一样，也是C++中重要的自定义数据类型。但是联合与结构体不同，结构体中的所有成员在任何时候都可以被访问，而联合中的成员在同一时刻只能有一个成员被访问。

链表是C++中一种重要的应用。链表有效地解决了动态数组问题，使得C++的数组更加灵活。

8.15 习题

一、填空题

1. 链表是一种复杂的数据结构，根据数据之间的关系可以把链表分成：_____、_____和_____等三种。

2. 结构体拥有_____的成员，是_____成员的集合。

3. 结构体变量的定义可以有3种方式：_____、_____和_____。

4. 结构体变量的引用格式为：_____。

5. 结构体在完成定义后便可以为它初始化，因为结构体中包含不同类型的变量，所以在为它初始化时也要＿＿＿＿＿＿＿＿的不同赋予不同类型的值。

6．若干个数据项按照一定的＿＿＿＿＿排列并通过＿＿＿＿＿链接起来，每个数据项都包含一个＿＿＿＿＿＿＿＿＿＿的指针，最后一个数据项没有指向固定目标的指针，其值为＿＿＿＿＿＿＿＿＿＿，这些链接到一起的数据项称为链表。

7．链表中的节点都包含一个＿＿＿＿＿＿＿＿＿＿的指针，因此链表中的数据在内存中的存储可以不必像数组中的数组元素一样是＿＿＿＿＿＿的。

8．向链表中插入节点可以分为：＿＿＿＿＿＿＿＿＿＿、＿＿＿＿＿＿＿＿＿＿和＿＿＿＿＿＿＿＿＿＿等3种情况。

9．联合与结构体有一点是不同的：在任一时刻，结构体中所有的成员都是＿＿＿＿＿＿＿＿＿＿的，而联合中的成员却＿＿＿＿＿＿＿＿＿＿是可以被访问的，其余的成员都是＿＿＿＿被访问的。

10．结构体中的每个成员都分配有自己独立的＿＿＿＿＿＿＿＿＿＿，而整个结构体的内存空间就是这个结构体中＿＿＿＿＿＿＿＿＿＿。

11．联合的内存空间是该联合的成员中＿＿＿＿＿＿＿＿＿＿的一个成员所占有的内存空间。

二、程序分析题

请认真阅读下面的程序，写出程序的运行结果，并上机验证。

1．代码如下：

```
#include <iostream.h>

struct st{
    char NO[10];
    float Tax;
};

main(){
    struct st st1;

    cout<<"NO=";
    cin>>st1.NO;
    cout<<"Tax=";
    cin>>st1.Tax;

    cout<<"NO="<<st1.NO<<endl;
    cout<<"Tax="<<st1.Tax<<endl;
}
```

2．代码如下：

```
#include <iostream.h>
```

```
union un{
    char NO[10];
    float Tax;
};


main(){
    union un un1;

    cout<<"NO=";
    cin>>un1.NO;
    cout<<"NO=";
    cin>>un1.NO;
    cout<<"你输入的NO="<<un1.NO<<endl;
    cout<<"Tax=";
    cin>>un1.Tax;

    cout<<"你输入的Tax="<<un1.Tax<<endl;
}
```

3. 代码如下：

```
#include <iostream.h>


struct myNode{
    int data;
    myNode *next ;
};


void createMyLink( myNode *&list){
    int data;
    myNode    *Node ;
    int i;

    for (i=9;i>=0;i--)  {
        Node = new myNode;
        Node->data =i;
        Node->next =list;
```

```
            list=Node;
        }
}


void outputMyLink( myNode  *&list){
    cout<<" 本链表结构如下 : "<<endl;
    myNode *p=list;

    while(p!=NULL){
        cout<<""<<p->data;
        p=p->next;
    }
    cout<<endl;
}


void findMyLink(myNode  *&list){
    int data;
    int i;

    cout<<" 请输入你要查找的数据 : ";
    cin>>data;
    i=0;
    myNode *p = list;

    while(p->data!=data)  {
        i++;
        p=p->next;
        if (p==NULL)  {
            break;
        }
    }


    if(p!=NULL)  {
        cout<<" 恭喜，你要查找的数据在本链表的第 "<<i+1<<" 个 "<<endl;;
    } else  {
        cout<<"Error:";
```

```
            cout<<" 很遗憾，没有找到你查找的数据！"<<endl;
        }
    }


int destroyMyLink(myNode * &list){
    char yN;

    cout<<" 警告：链表将被销毁，是否继续（Y/N）？";
    cin>>yN;
    if(yN=='n'||yN=='N')   {
        return 0;
    }


    while(list!=NULL)   {
        myNode *p =list;
        list = list ->next;
        delete p;
    }
}

main(){
    cout<<"*********************************************"<<"\n";
    cout<<"*                    链表的应用              *"<<"\n";
    cout<<"*********************************************"<<"\n";
    myNode * head=NULL;

    createMyLink(head);
    findMyLink(head);
    outputMyLink(head);
    destroyMyLink(head);
    outputMyLink(head);
    cout<<"*********************************************"<<"\n";
}
```

三、问答题

1. 什么是结构体？它与数组有什么不同？

2．结构体的定义有几种方式？请分别举例说明。

3．什么是链表？链表分为哪几种？

4．什么是联合？联合与结构体有什么不同？它们在内存分配上有什么区别？

第9章　输入/输出流

本章学习目标

↘ **C++流的概念**　　↘ **熟练运用cin流**

↘ **熟练运用cout流**

如果你有过C语言的使用经验，那么就应该知道C语言提供的标准I/O库stdio，这个标准库定义了面向控制台（显示器、键盘）的输入和输出。而C++则提供了一个标准的I/O库iostream，准确地讲应该叫I/O流库。iostream是面向对象的，它可以实现C语言中stdio库的所有功能。而与stdio相比，iostream更安全、更有效。要使用iostream，必须在程序文件的开始包含iostream库，这在前面章节中已经使用过很多次了，其格式为：

```
#include <iostream.h>
```

在前面章节的程序中使用的输入流是cin，输出流是cout。cin和cout都是iostream提供的全局流对象。

本章的目的是使读者能够全面、准确地理解C++中输入/输出流的概念、意义和应用。

9.1　输入/输出流概述

所谓流就是一个字节的序列。在C++中，I/O操作是通过流来实现的。事实上，C++本身并不提供I/O操作，C++的I/O操作是通过C++标准库来实现的。

iostream类库是C++的输入/输出库，其中iostream.h头文件又是这个类库中最重要的文件，它提供了所有I/O操作所需要的基本信息。iostream.h头文件中包括了前面所讨论过的最常用到的cin和cout对象。

此外，在iostream.h头文件中还包含cerr和clog对象，表9-1列出了C++提供的4种标准流。

表9-1　iostream类库中定义的标准流

流　名	含　义	隐含设备
cin	标准输入	键盘
cout	标准输出	屏幕
cerr	标准出错输出	屏幕
clog	cerr的缓冲形式	屏幕

除了iostream.h头文件外，iostream类库还提供了以下重要的头文件：

* iomanip.h　包含带参数的流操作。

* fstream.h　对文件I/O的操作。

- stdiostream.h 混合使用C/C++风格的I/O操作的相关信息。
- strstream.h 包含对内存格式化I/O操作的相关信息。

9.2 cout流

使用C++的cout流时需要配合运算符<<一起应用。运算符<<又叫插入运算符，因为<<是将其后面所跟的数据插入到它前面的数据流中。下面通过一个例子来看一下C++的cout流的应用。

实例9-1 使用cout流。

```
1 :#include <iostream.h>
2 :
3 :main(){
4 :    int i;
5 :
6 :    i=10;
7 :    cout<<"Hello World!"<<endl;;
8 :    cout<<100<<endl;
9 :    cout<<i<<endl;
10 :}
```

程序运行结果如下：

```
Hello World!
100
10
```

在这个例子中，将cout流和插入运算符配合使用来输出结果。具体解释如下：

（1）程序的第7行是将字符串"Hello World!"通过插入运算符<<插入到cout流中输出。

（2）程序的第8行是将数值100通过插入运算符<<插入到cout流中输出。

（3）程序的第9行是将变量i的值通过插入运算符<<插入到cout流中输出。

插入运算符可以被连续使用多次，例如：

```
cout<<"Hello"<<" "<<"World!";
```

这条语句的运行结果同样是输出字符串"Hello World!"。

9.3 cin流

C++的cin流需要配合运算符>>一起应用，表示将>>后面跟的值读取并存储，其格式如下：

```
cin>> 变量;
```

>>运算符也可以被连续使用，例如：

```
cin>> 变量1>> 变量2>> 变量3 … >> 变量n;
```

需要注意的是，cin流在接受用户输入时，以回车符作为输入内容的结束符，因此只有在用户输入完成后按下回车键，用户的输入才会被cin流处理。

实例9-2 使用cin流。

```cpp
#include <iostream.h>


main(){
    int i;

    cout<<"i=";
    /* 流输入 */
    cin>>i;
    /* 清空屏幕 */
    system("cls");
    cout<<" 你输入的值是 : "<<"i="<<i<<endl;
    /* 暂停 */
    system("pause");
    system("cls");

}
```

9.4 上机操作

实例9-3 使用cin和cout输入并倒序输出一列数值。

```cpp
#include <iostream.h>


void Sort();


void Sort(){
    int i,j;
    float t;
    float a[5];


    for (i=0;i<5;i++)  {
        /* 利用 cout 流输出提示 */
        cout<<"a["<<i<<"]=";
        /* 利用 cin 流向数组中写入数据 */
        cin>>a[i];

    }
```

```
    for (i=0;i<5;i++)
        for (j=i+1;j<5;j++)
            if (a[i]<=a[j])  {
                t=a[i];
                a[i]=a[j];
                a[j]=t;
            }
    cout<<endl;
    cout<<" 排序结果 : ";
    for (i=0;i<5;i++)
      if (i!=4)
        cout<<a[i]<<"->";
      else if (i==4)
        cout<<a[i]<<endl;
}


main(){
    cout<<"***********************************************"<<"\n";
    cout<<"*                  cout 流与 cin 流                *"<<"\n";
    cout<<"***********************************************"<<"\n";
    Sort();
    cout<<"***********************************************"<<"\n";
}
```

程序运行结果如下：

```
***********************************************
*                  cout 流与 cin 流                *
***********************************************
a[0]=1
a[1]=2
a[2]=3
a[3]=4
a[4]=5


排序结果 : 5->4->3->2->1

***********************************************
```

9.5 小结

　　C++没有内置的输入/输出功能，这不是C++的缺点，恰恰相反，这正是它的优点。因为这样可以脱离系统硬件，不必为特定的操作系统定义输入/输出。C++的输入/输出是通过输入/输出流来实现的，即通过include载入C++的标准输入/输出流iostream，从而实现输入/输出操作。本章结合实例讲解了iostream流的基本输入/输出操作——cin和cout，请读者仔细阅读本章实例，深入理解cin和cout的使用。

9.6 习题

一、填空题

　　1．C++中提供了一个类库，它能实现类似C语言中的I/O标准库stdio的全部功能，并且这个标准库是面向对象的，这个标准库是＿＿＿＿＿＿＿。

　　2．cout流需要配合＿＿＿＿使用，cin流需要配合＿＿＿＿使用。

二、问答题

　　在iostream.h头文件中定义的4个基本输入/输出流是什么？分别代表什么意义？

第10章 异常处理与错误

本章学习目标

▶ 了解为何需要异常处理 ▶ 熟练掌握throw 表达式

▶ 熟练掌握try catch表达式 ▶ 了解如何进行程序调试

　　本章的目标是对C++中的异常有一定的了解，注意几种常见的错误。异常处理是对程序中可能出现的错误给出相应的解决办法，使程序不至于崩溃。在编写程序的过程中，难免会出现错误，如程序只接受特定类型的输入，而其他输入却可能造成程序错误，这时就要在程序中加上处理错误输入时相应的解决方法，也就是异常处理。这就像人们在做事的过程中对事情可能发生的不良后果做好心理准备一样，不至于在程序遇到错误时就导致程序崩溃。

10.1 为何需要异常处理

　　在编写程序时，程序会出现各种各样的错误。当一个程序出错时，程序本身一般是不能自动处理的，但是如果有异常处理功能的话，那么程序即使遇到错误，也会对错误进行处理，不至于因为某些错误而使程序崩溃。当然处理程序的错误是相当困难的，有些错误可能不易发现，因而在发现错误后就要更正它，如果有些地方存在反常行为，则应给出相应的处理。

　　总之，异常就是程序运行时出现了不正常的现象。例如，可能耗尽了内存，也可能遇到了非法输入。当遇到这些不正常的现象时，就要使程序立即进行处理，但是异常处理并不存在于程序的正常功能之内，只有遇到不正常现象时才会使用异常处理。

　　要想设计一个好的程序，就要使程序具备异常处理的能力，当正常的程序代码不能处理时，就要使用异常处理程序来对出现的不正常现象给出相应的处理。当遇到问题时，程序会转到处理不正常现象的异常部分进行处理，检查出相应的错误，并指出程序出现了什么样的问题，向用户提供消息，说明出现了什么样的错误。

　　大家在玩游戏时也许遇到过这样的情况，游戏过程中突然弹出一个对话框，说游戏错误，是否需要调试，这也就是游戏出现了不正常现象，此时只能退出游戏。

　　C++的异常处理包含三部分的内容：throw表达式、try块与catch子句。本章将对这三部分进行讲解。首先看下面的代码：

```
#include <iostream>
using namespace std;
int division(int a,int b)
```

```
{
    return a/b;
}

int main()
{

    int n1,n2;
    cout<<" 请输入两个数 : "<<endl;
    cin>>n1>>n2;                    // 输入 n1 与 n2 的值
    cout<<" 相除后的值为 : "<<division(n1,n2)<<endl;          // 输出相除后的结果
    system("pause");
    return 0;

}
```

如果执行这个程序，那么在输入数1与0时，这个程序将出现错误，而不会有任何输出。并在程序中提示哪个地方出现了错误。这是因为不能把0作为除数，所以这时就要使用异常处理机制。

10.2 throw表达式

前面已经了解到，程序可能存在着这样或那样的错误，当遇到错误时，就要给出相应的解决办法，使程序不至于产生错误而崩溃，这时便可以使用throw表达式。throw表达式可以用来说明程序遇到了什么样的错误，系统通过throw表达式抛出异常。throw表达式的格式如下：

```
throw 表达式 ;
```
首先是throw关键字，然后再使用一个表达式，这样就组成了throw表达式。

上一节中的错误当然也可以使用前面学习过的标准错误来进行说明，修改上一节中的程序，如下所示：

```
#include <iostream>
using namespace std;
int division(int a,int b)
{

    return a/b;

}
int main()
{

    int n1,n2;
    cout<<" 请输入两个数 : "<<endl;
    cin>>n1>>n2;                                    // 输入 n1 与 n2 的值
```

```
    if(n2==0)                                           // 判断 n2 的值是否为 0
    {
        cerr<<"n2 的值不能为 0！ "<<endl;
        system("pause");                                // 暂停程序
        return 0;                                       // 函数的返回值
    }
    cout<<" 相除后的值为 : "<<division(n1,n2)<<endl;     // 输出相除后的结果
    system("pause");
    return 0;
}
```

上面的程序如果再遇到 n2 为 0 的输入时，将会输出 "n2 的值不能为 0！ " 的提示信息，然后退出程序，而不会再执行程序后面的代码。如果使用异常呢？上面的程序又将会如何？下面就使用异常来改写上面的程序，如下所示：

```
#include <iostream>
#include <stdexcept>                                    // 使用命名空间 stdexcept
using namespace std;
int division(int a,int b)                               // 求两个数相除
{
    return a/b;                                         // 返回相除后的结果
}
int main()
{
    int n1,n2;
    cout<<" 请输入两个数 : "<<endl;
    cin>>n1>>n2;                                        // 输入 n1 与 n2 的值
    if(n2==0)                                           // 判断 n2 的值是否为 0
    {
        throw runtime_error("n2 的值不能为 0！ ");
        system("pause");                                // 暂停程序
        return 0;                                       // 函数的返回值
    }
    cout<<" 相除后的值为 : "<<division(n1,n2)<<endl;     // 输出相除后的结果
    system("pause");
    return 0;
}
```

这里就使用了 throw 来抛出一个异常，说明当 n2 的值为 0 时，程序会出现错误。runtime_error

类型为标准库stdexcept类中的一种，所以在使用时一定要注意包含相应的头文件。关于它的使用将在下一节给出。

10.3 try块与catch的使用

前面已经介绍了可以使用throw表达式来抛出程序中存在的错误，可是抛出错误后程序该怎么处理这个错误呢？这时就要使用try块与catch子句。

错误处理部分便由catch部分来处理。在try块中，如果程序有错的话，会抛出一个异常，然后抛出的异常会被catch子句处理，而且通常会有多个catch子句，try块的定义如下所示：

```
try
{
  // 条件
}
catch(异常1)
{
  // 相应的处理
}
catch(异常2)
{
  // 相应的处理
}
…
catch(异常n)
{
  // 相应的处理
}
```

从定义中可以看出，try块是以关键字try开头的，后面接的是处理的条件，然后是catch子句，当然这个catch子句可以有任意多个，catch子句以关键字catch开头，后面接的是相应的异常，然后是相应的异常处理方法。catch子句只会执行与抛出条件中的异常相符的子句，执行完后将执行最后一个catch子句后面的语句。下面就改写上一节中的程序。

实例10-1 try语句的使用。
代码如下：

```
#include <iostream>
#include <stdexcept>
using namespace std;
int division(int a,int b)                          // 使两个参数进行 "/" 运算
```

```
{
    return a/b;
}

int main()
{
    int n1,n2;
    cout<<" 请输入两个数："<<endl;
    cin>>n1>>n2;                                            // 输出 n1 与 n2 的值
    try                                                    //try 块
    {
        if(n2==0) throw runtime_error("n2 的值不能为 0：");
    }
    catch(runtime_error err)                               //catch 子句
    {
        cout<<err.what()<<endl;                            // 抛出的异常信息
        system("pause");
        return 0;
    }
    cout<<" 相除后的值为："<<division(n1,n2)<<endl;    // 输出相除后的结果
    system("pause");
    return 0;
}
```

（3）运行程序，将出现如图10-1所示的窗口。

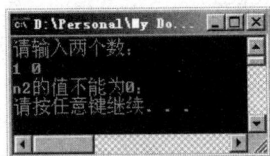

图10-1　输出提示信息

程序分析

本程序使用了try块，如果n2的值为0，那么将由throw抛出一个异常runtime_error，catch子句将捕获这个异常，然后输出异常的信息。

程序中首先定义了一个int类型的函数division，它有两个int型的形参，函数体中返回两个形参求"/"后的结果。

在主函数中，定义了两个int型变量n1与n2，作为函数的实参，在获取n1与n2的值后，将使用try块测试n2的值，如果为0则会抛出一个异常，否则正常调用函数。在输出信息时使用了err.what()的返回值作为抛出的异常信息。what()是runtime_error类的一个成员函数，它是一个无参函数，返回C风格字符串，用于初始化runtime_error的string对象的副本。

10.4 常见错误

在使用C++编写程序时，难免会出现各种各样的错误，而对于编程新手来说，出现错误的概率更大。许多新手在刚学习编程时遇到的错误往往是相同的，本节的内容就是对这些常见的错误进行汇总，让大家在开始编程的时候便注意这些问题，养成良好的编程习惯。

（1）不包含相应的头文件。这是一个经常犯的错误，例如下面的代码：

```
int main()
{
    cout<<"Hello world!"<<endl;
    return 0;
}
```

这里就没有包含相应的输入输出头文件，编译上面的代码将会出现下面的错误：

```
In function "int main()":
error:'cout' undeclared (first use this function)
error: (Each undeclared identifier is reported only once for each function it
appears in.)
error: 'endl' undeclared (first use this function)
```

其中的cout与endl就是因为没有包含相应的头文件而导致它们在使用的过程中未定义。

（2）主函数定义为void型，如下所示：

```
#include <iostream>
using namespace std;
void main()
{
    cout<<"Hello world!"<<endl;
}
```

编译程序后，会出现如下的错误：

```
'main' must return 'int'
```

说明主函数必须返回int型，当然这是在标准C++里面定义的，可能在使用其他的编译器时，使用void类型作为返回类型时，程序也能正常执行，但是这并不能保证在所有的编译器里面都能通过。

（3）在写语句时漏掉语句后面的分号，这个错误在开始写程序时会经常出现。如下面定义两个变量：

```
int a=3;
int b=5
```

定义变量a没有错误，可是在定义变量b时由于没有写分号，将导致如下的错误：

```
'main' must return'int'
error: expected',' or ';' before "return"
```

错误提示在return前面可能少了一个";"，所以在开始编写程序时一定要注意这一点。

（4）在判断条件中把"="当成"=="，看下面的程序错在哪里。

实例10-2　错把操作符"="当成"=="。

代码如下：

```cpp
#include<iostream>
using namespace std;
int main()
{
    int a;
    cout<<" 请输入 a 的值 !"<<endl;
    cin>>a;                          // 输入 a 的值
    if(a=5)                          // 给变量 a 赋值为 5
    {
        cout<<"a 的值为 5 ！ "<<endl;
    }
    else                                  // 其他条件
    {
        cout<<"a 的值不为 5 ！ "<<endl;
    }
    system("pause");
    return 0;
}
```

运行程序，将出现如图10-2所示的窗口。

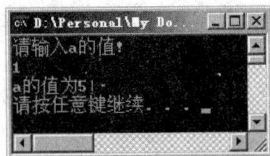

图10-2　输出a的值是否为5

程序分析

在上面的程序中，当输入1时也输出了"a的值为5！"，这是为什么呢？原来在if语句中错误地把"="当作了"=="，if语句的条件变为把a赋值为5，最后a的值为5，是一个非0数值，只要赋的值不为0，那么if条件都为真，if语句都将被执行。所以将上面的程序修改为如下程序，运行结果才是正确的。

```cpp
#include <iostream>
using namespace std;
int main()
```

```
    {
        int a;
        cout<<" 请输入 a 的值！"<<endl;
        cin>>a;                          // 输入 a 的值
        if(a==5)                              // 判断 a 的值是否为 5
        {
            cout<<"a 的值为 5！ "<<endl;
        }
        else                           // 其他条件
        {
            cout<<"a 的值不为 5！ "<<endl;
        }
        system("pause");
        return 0;
    }
```

（5）使用了未定义的变量。C++中如果一个变量没有定义，那么它是不能使用的。如下面的代码：

```
#include <iostream>
using namespace std;
int main()
{
    int a=1;
    int b=2;
    c=a+b;
    cout<<"a 与 b 的和为："<<c<<endl;
    system("pause");
    return 0;
}
```

编译上面的代码将会出现如下编译错误：

```
In function 'int main()':
error: 'c' undeclared (first use this function)
error: (Each undeclared identifier is reported only once for each function it appears in.)
```

这里的错误就是因为在用变量c时没有进行定义造成的。

（6）在if与for语句中多使用了分号，如下所示：

```
if(a==0);
```

```
{
    cout<<"a 的值为 0!"<<endl;
}
```

上面的if语句中，在条件后面加了一个分号，所以不论a的值是否为0，那么括号里面的语句都将被执行。在for语句的使用过程中也有一点需要注意，请看下面的代码：

```
for(int i=0;i<5;i++);
{
    cout<<" 数组中包含元素："<<a[i]<<endl;
}
```

上面的for语句在条件后面加了一个分号，所以for语句什么事也没有做，接着再执行后面的语句，这时数组将没有定义，因为i的作用域只在for语句中。

10.5 程序调试

在程序出现错误后就要对这个程序进行检查，检查程序存在什么样的错误，当然有些错误是不能一眼就看出来的，这时就要通过调试，也就是说，对程序的具体执行过程进行了解。通过调试可以发现程序在执行过程中，变量是如何变化的，比较变量的变化是否与预期的一样，如果不一样则可能发生了错误，那怎样调试程序呢？这便是本节要学习的内容。

10.5.1 设置断点

如果要查看某程序段在整个程序中是什么时候执行的，或执行了多少次，这时就可以在这段程序中加上断点，加入断点之后，每当程序运行到这个程序段时都会返回程序的代码窗口，以高亮显示断点处的代码正在被执行。

设置断点的方法有两种：一种是直接设置断点，另一种是通过菜单项设置断点。而直接设置断点因为使用方便，所以是程序员们经常使用的一种方法，下面就对这两种方法进行讲解。

1.通过菜单项设置断点

（1）把光标移动到要设置断点的代码段上，如图10-3所示。

（2）单击菜单栏的"调试"按钮，在弹出菜单中选择"切换断点"命令，如图10-4所示。

（3）这时相应的代码段前就插入了一个断点符号，如图10-5所示。

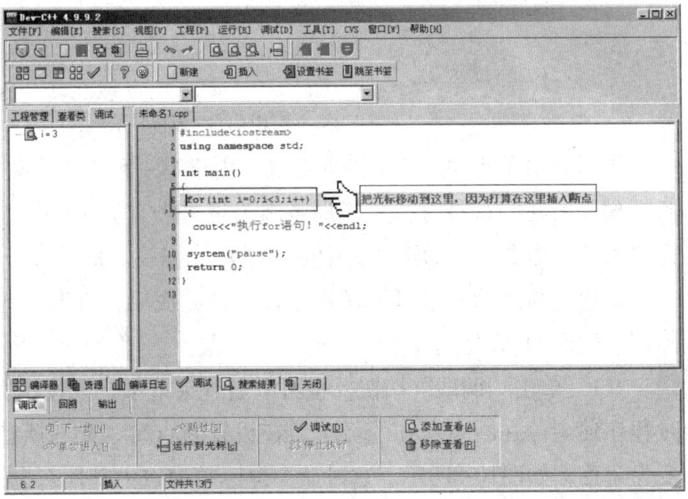

图10-3　把光标移动到要设置断点的所在行

这样，在开始准备设置断点的地方就设置好一个断点。

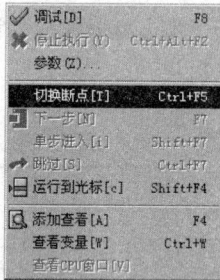

图10-4　选择切换断点　　　　　　　　图10-5　插入断点后所在行的变化

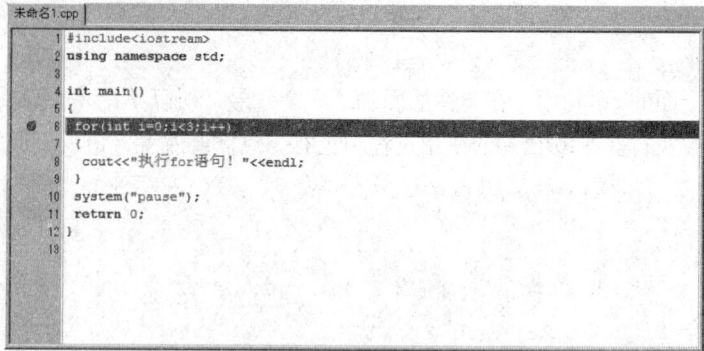

2. 直接插入断点

（1）将鼠标指针移动到要插入断点所在行的行号处，如图10-6所示。

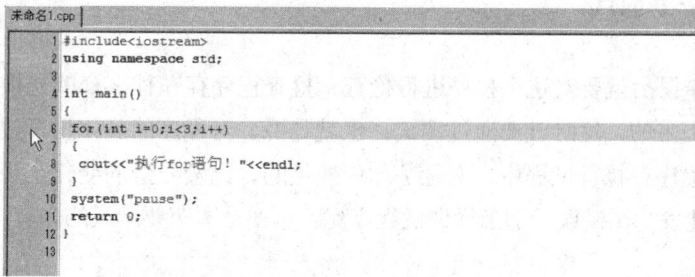

图10-6　直接在要插入断点的行号上单击

（2）在所在行的行号上单击鼠标，即可产生一个断点，如图10-7所示。

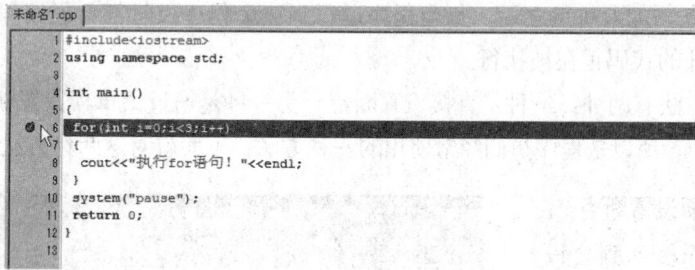

图10-7　单击行号后所在行的变化

上面使用了两种方法来设置断点，由于第一种方法比较方便，所以推荐使用第一种方法。

如果所在的行没有行号，那是因为行号没有显示出来。这时可以通过选择"工具"→"编译器属性"命令，然后选择其中的"显示"标签，把"行号"前的复选框选中便可。

设置完断点后，就可编译程序，编译完成后，再调试程序，这时程序就会在所在的断点处停留，过程如下：

（1）单击"调试"按钮，也可以通过菜单项，或按快捷图标，或按F8快捷键进入调试，这时程序便会停在断点所在的行上。在设置断点时，光标所在行以红色显示，而这时断点所在行变为蓝色，如图10-8所示。这时命令提示符窗口中却什么也没有显示，如图10-9所示。

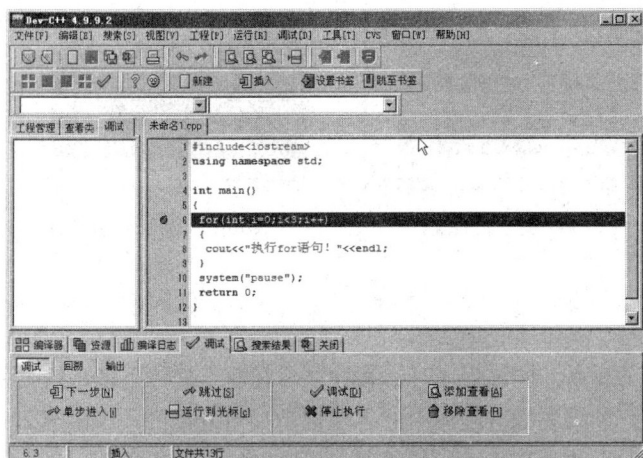

图10-8 调试后所在行的变化

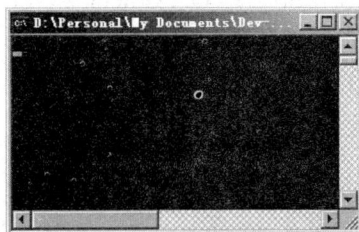

图10-9 命令提示窗口没有任何输出

(2) 单击"下一步"图标，则编译器窗口中以蓝色显示程序执行到了哪个地方，如图10-10所示。

当程序运行到for语句里时，再单击"下一步"图标，命令提示符窗口变为如图10-11所示。

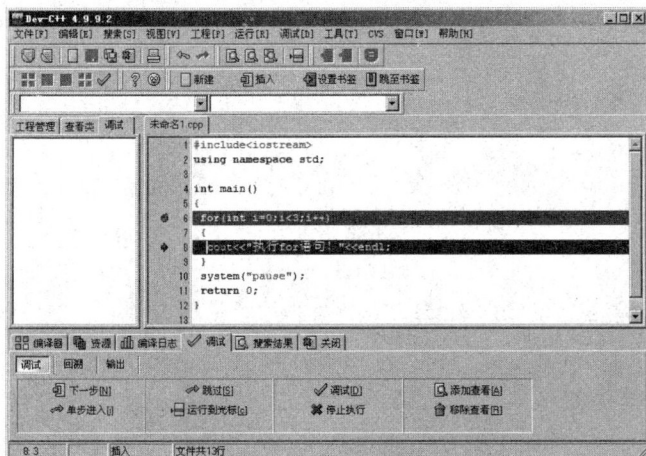

图10-10 单击"下一步"图标后程序所在行的变化

图10-11 命令提示符窗口

编译器窗口如图10-12所示。

一直单击"下一步"图标，在编译器中，高亮显示的语句将在for语句的条件与执行语句之前反复执行，直到程序退出for语句，而命令提示符如图10-13~图10-15所示。

直至程序运行到最后一个语句，然后高亮显示，最后单击"停止执行"图标退出程序。

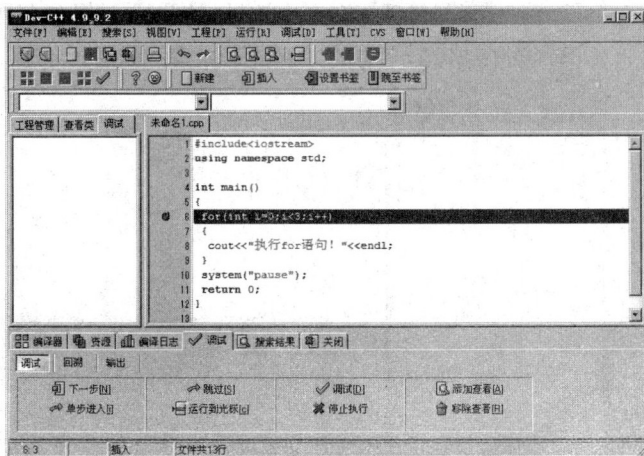

图10-12 代码所在行的变化

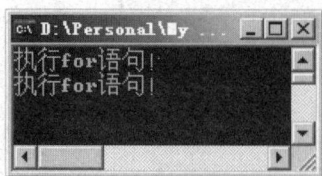

图10-13　第二次的输出　　　　　图10-14　第三次的输出　　　　　图10-15　第四次的输出

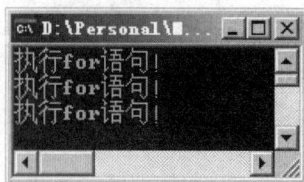

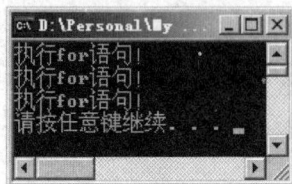

10.5.2　添加查看

前面已经在程序的源代码中设置了断点，调试程序时，程序便会在断点处停下来，这时便可以在断点所在的地方查看变量的值。如果要知道程序中的各变量是如何变化的，应该怎么办呢？请看下面的示例。

实例10-3　通过"添加查看"查看变量的变化。

可以在程序中添加断点之后再添加监视点。具体过程如下：

（1）新建源文件"eg1003.cpp"，输入下面的代码，然后在for语句的条件所在行设置断点。

```cpp
#include <iostream>
using namespace std;
int main()
{
    int a[5]={1,2,3,4,5};
    for(int i=0;i<5;i++)   // for 语句完成数组元素的输出
    {
        cout<<" 数组 a 中含有元素 : "<<a[i]<<endl;
    }
    system("pause");
    return 0;
}
```

（2）在"工程管理"命令中选择"调试"选项，如图10-16所示。

（3）单击编译器下方的"调试"标签，如图10-17所示。

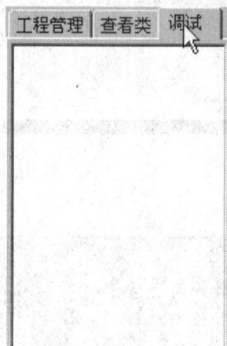

图10-16　调试窗口

图10-17　"调试"标签

（4）单击"调试"图标，开始调试程序。

（5）单击"添加查看"图标，可以直接在编译器下方单击图标；也可以选择"调试"菜单，再选择"添加查看"命令；还可以通过按F4快捷键，如图10-18所示。

（6）在弹出的"新变量"窗口中输入变量名，这里输入i，然后把光标移动到for循环体语句上，再次单击"添加查看"图标，在弹出的新变量窗口中输入a[i]。这时调试窗口中出现下列变量，如图10-19所示。

（7）一直单击"下一步"图标，这时调试窗口中的变量值不断在改变，图10-20~图10-24分别列举了每次单击"下一步"图标后变量的值的变化情况。

图10-18　通过下拉菜单选择"添加查看"　图10-19　"添加查看"后调试窗口中的变量　图10-20　第一次变量的变化

（8）单击"停止执行"，如图10-10所示。

在程序中设置的添加查看的变量，其值的变化情况为：i的值从0~4逐步递增，而a[i]的值从1~5逐步递增。

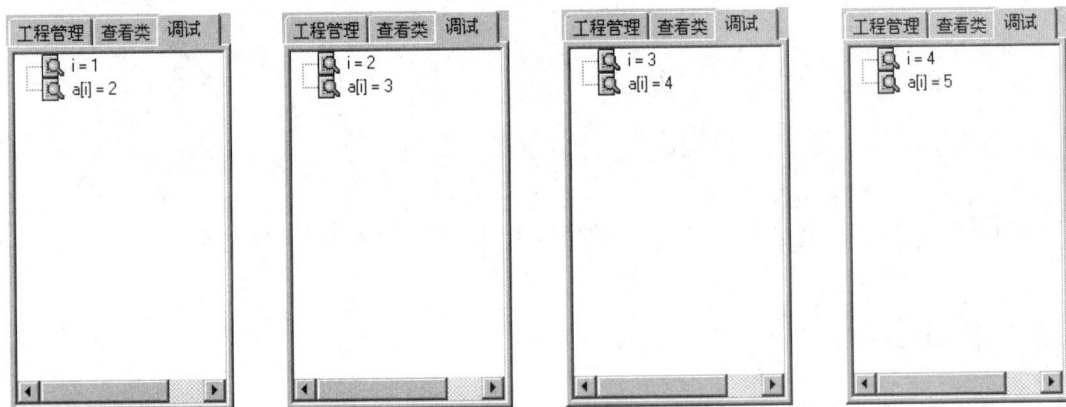

图10-21　第二次变量的变化　图10-22　第三次变量的变化　图10-23　第四次变量的变化　图10-24　第五次变量的变化

10.6　上机操作

　　实例10-4　从键盘上输入两个整数a、b，求其商。如果除数为零，则不进行运算，在屏幕上输出"除数不能为零"的异常信息，否则输出正确结果。代码如下：

```
#include <iostream>
Void Div(int a,int b)
{
    int c;
    try
    {
        if(b==0) throw b;
      c=a/b;
      cout<<"a/b="<<c<<endl;
    }
    catch(int z)
    {
        cout("除数不能为零！")<<endl;
    }
}
void main()
{
    int a,b;
    cout<<"请输入被除数 a 和除数 b 的值："<<endl;
    cout<<"a=";
    cin>>a;
    cout<<"b=";
    cin>>b;
    Div(a,b);
}
```

运行程序，输入数值分别为a=6、b=2和a=3、b=0，查看每次程序运行的结果。

10.7 小结

本章首先介绍了为什么会有异常，以及处理异常的三个步骤。最后介绍了在程序中出现错误时该如何去进行调试。

10.8 习题

一．填空题

1. C++的异常处理包含的三个部分内容是_____。

2．throw表达式的格式是_____。

3．错误处理部分由_____部分来处理。

二．编程题

1．设置断点进行程序的调试。

2．分析下面程序的运行结果。

```cpp
#include <iostream>
using namespace std;
int main()
{
    try{
        try{
            throw'a';
            cout<<"first!"<<endl;
            }
        catch(char)
        {
            throw;
            cout<<"second!"<<endl;
        }
    }
    catch(…)
    {
        cout<<"抛出异常！"<<endl;
    }
    return 0;
}
```

参考文献

[1] 普拉达. C++ Primer Plus[M]. 张海龙, 袁国忠, 译. 6版. 北京: 人民邮电出版社. 2020

[2] 谢丙堃. 现代C++语言核心特性解析[M]. 北京: 人民邮电出版社. 2021

[3] 钱林松, 张延清. C++反汇编与逆向分析技术揭秘[M]. 2版. 北京: 机械工业出版社. 2021

[4] 苏小红, 蒋远, 单丽莉, 等. 程序设计实践教程: C++语言版[M]. 北京: 机械工业出版社. 2021

[5] 谭浩强, C++程序设计[M]. 4版. 北京: 清华大学出版社. 2021

[6] 肖连, 从零开始 C++程序设计基础教程[M]. 北京: 人民邮电出版社. 2021

[7] 董兴业, 瞿有利, 王涛. C++面向对象程序设计[M]. 北京: 清华大学出版社. 2021

[8] 张琨, 张宏, 朱保平. 数据结构与算法分析: C++语言版[M]. 北京: 人民邮电出版社. 2021

[9] 张远龙. C++服务器开发精髓[M]. 北京: 电子工业出版社. 2021

[10] 向志华, 张莉敏, 邓怡辰, 等. C++程序设计[M]. 北京: 清华大学出版社. 2021